CUSTOM-MADE OFFICE 3

Highlighting Exclusive Corporate Culture and Temperament

凸显专属的企业文化气质

定制办公 3

Shenzhen Design Vision Cultural Dissemination Co., Ltd
深圳视界文化传播有限公司 编

辽宁科学技术出版社
LIAONING SCIENCE AND TECHNOLOGY PUBLISHING HOUSE

图书在版编目（CIP）数据

定制办公．3／深圳视界文化传播有限公司编．--
沈阳：辽宁科学技术出版社，2015.7
 ISBN 978-7-5381-9311-4

Ⅰ．①定… Ⅱ．①深… Ⅲ．①办公室－室内装饰设计
－作品集－世界－现代 Ⅳ．① TU243

中国版本图书馆CIP数据核字（2015）第 157007 号

策划制作：深圳视界文化传播有限公司（www.dvip-sz.com）
总 策 划：蒙俊伶
编　　辑：陶心璐
翻　　译：赵　耀　曹　鑫
装帧设计：潘如清
联系电话：0755-82834960／0755-83456359

出版发行：辽宁科学技术出版社
（地址：沈阳市和平区十一纬路29号　邮编：110003）
印 刷 者：深圳市新视线印务有限公司
经 销 者：各地新华书店
幅面尺寸：245mm×335mm
印　　张：40
插　　页：4
字　　数：300千字
出版时间：2015年7月第1版
印刷时间：2015年7月第1次印刷
责任编辑：王玉宝　视界
责任校对：合力

书　　号：ISBN 978-7-5381-9311-4
定　　价：380.00元(USD 80.00)

联系电话：024—23284376
邮购热线：024—23284502
E-mail:lnkjc@126.com
http://www.lnkj.com.cn

PREFACE 1
序言 1

Office means more than just a working space
办公室不仅仅是一个工作空间

Many tempers come together in one area called "office". Dozens if not hundreds of people with different ambitions and desires come here, and one of the main tasks an architect and a client work on is to do their best to ensure that a job will be done in a comfortable environment. The more convenient the working space, the better the time employees have there, the better the output of their work is.

As architects we find it very appealing that companies go for individuality. Now every project has become an opportunity to create a one-of-a-kind space that reflects a company's business, becomes its brand identity and the best possible working place for a team. More and more companies recognize that apart from a workplace, meeting rooms and utility rooms their staff need places for so called random meetings. Places where an interesting idea may come up, and where impromptu brainstorms will take place. Informal setting of such places allows one to unwind and catch a break and then proceed to work with renewed vigour. With increasing frequency we create offices with game zones, porches, several coffee-points, and small spaces for phone conversations or meetings, where employees can contemplate in peace and unwind or vice versa—discuss things on the fly. It is more common now that we start out from the needs of teams, and we see more often that businesses care for their staff as much as for their clients. This trend allows creation of interesting creative spaces, unlike unchanging offices of the past. We are curious to know what the years to come will bring, but for some of our clients the future has already come today!

许多有着不同性情的人聚集在一个被称为"办公室"的地方，设计师与其客户的主要任务就是尽其所能确保工作在舒适的环境中完成。工作环境越便利，员工就过得越愉快，工作成果也就越好。

作为设计师，我们发现公司都追求个性，这引起了我们极大的兴趣。现在每一个项目都已经成为我们创造独一无二的空间的机遇，这样的空间反映公司业务，同时成为公司的品牌标识和工作团队的最佳工作地点。越来越多的公司认识到，除了工作室、会议室和公共设施区域，员工需要空间来开展所谓的随机会议。将会出现一些可以引发有趣想法和即兴头脑风暴的空间。这类空间的随意设置让人们能够放松和休息，从而精力充沛地工作。我们更频繁地在办公室内创建游戏区、走廊、咖啡点、小型电话室或会议室，员工们可以在这些地方安静地思考、放松或讨论正在忙碌的事情。从团队的需求出发已经变得越来越普遍，我们更多地看到企业对员工的关心就像他们对客户的关心一样。这一趋势推动了创建有趣的创意空间，不像过去那些一成不变的办公室。我们想知道未来会发生什么，但是对于我们的一些客户来说，未来已经到来！

Amir Idiatulin,
CEO IND Architects, IND Office

阿米尔·艾迪尔图林
IND建筑事务所、IND工作室执行总裁

www.indarchitects.ru | www.indoffice.ru
www.facebook.com/indarchitects | www.facebook.com/INDoffice

PREFACE 2
序言 2

The importance of building brand identity in interior design
论室内设计中塑造品牌标识的重要性

Globalization means more and faster flow of information, with greater opportunities, but also more competition. Using brand identity is an important tool, to attract new clients and the very best talents. A strong brand generates higher profit, better margins, and a longer loyalty from customers and from employees. The brand identity of an office is crucial to create new corporate values.

全球化意味着更多且更快速的信息流动,伴随着更大的机会,也带来更多的竞争。使用品牌标识是一个很重要的手段,以此可以吸引新客户和最优秀的人才。一个强大的品牌能够创造出更高的效益和更高的利润,同时获取顾客和员工的更长久的忠诚和信赖。办公室的品牌标识对创造新的企业价值观尤为重要。

Where do you like to work?
您想在哪儿工作呢?

We live in a rapidly changing world. Technology gets smarter, smaller and more flexible. We can work almost anywhere. In order to work you actually do not have to go to work. The anatomy of work is not where you work, it is what you do. The office is not where you store your things. No, the office is a place to discuss, report, socialize and to make new ideas grow. The office has become more "we-oriented" rather than "me-oriented". To unite a company and employees, it is important to focus on new ways of collaboration, and to build a strong brand identity that everyone can relate to.

我们生活在一个快速变化的世界。技术更加智能、精巧和灵活。我们可以随处工作。实际上你不必去上班也可以工作。工作的意义不在于你在哪工作,而在于你做什么。办公室并非你储存东西的地方,而是可以讨论、作报告、社交和开拓新思维的地方。办公室变得越来越"以我们为中心"而非"以我为中心"。为了团结公司和员工,注重新的合作方式以及建立一个能够与每个人都产生联系的强大的品牌标识是极其重要的。

To illustrate what a strong brand identity could be, I have selected three examples from our office projects, all of them are located in Stockholm, Sweden. You can find more pictures and drawings of the Skype and Chimney Group projects in this book.

为了说明一个强大的品牌标识是怎样的,我从我们的办公室项目中选了三个位于瑞典斯德哥尔摩的案例。你可以在本书中找到更多有关于Skype办公室和Chimney办公室的照片和图纸。

1) Coop concept
Coop办公室理念

Coop Sweden´s HQ was relocated from an A+ to a B- location in Stockholm. One major challenge was to create a brand identity with a "home feeling" that could unite the staff of app. 650 people. The colors of the membership card, and the circles in the logo, where symbolized in the main atrium as colored stripes and three round light fittings.

Coop集团瑞典总部从非常好的地段搬迁至斯德哥尔摩较差的地段。主要的挑战就是需要创建一个具有"家的感觉"的品牌标识,从而使近650名员工团结在一起。会员卡的颜色和商标的外光圈在主厅以彩色条纹和三圈灯饰展示出来。

2) Skype concept
Skype办公室理念

The idea behind the Skype software has generated the brand design in the Stockholm office. Skype is a useful and playful tool that allows people to connect and make video calls all around the world. The contemporary interior design not only creates a good looking environment, but also an inspiring work place, that allows good, crazy, and brilliant ideas to develop. The cloud logo and electronic gears are symbolized in the lighting fixtures and in the wall paper design.

在Skype软件的理念上产生了对斯德哥尔摩办公室的品牌设计。Skype是一个有用且有趣的工具，人们可以互相联系并且在世界各地可以进行视频通话。当代室内设计不仅创建了一个漂亮的环境，更是一个令人鼓舞的工作场所，激发出优秀的、疯狂的和杰出的创意。云形商标和电子齿轮体现在照明灯具和墙纸设计上。

3) Chimney Group concept
Chimney集团理念

Chimney Studios are storytellers at heart and use film and communication expertise to produce over 6,000 outputs for more than 60 countries every year. The work covers each step of the creation process, from conception through development, production and world-class post-production.
The design concept is The Chimney Family – welcoming, professional and sporting gifts and souvenirs from friends all over the world. The interior is a mix of second hand and new furniture, all held together by a toned down color scheme and soft fabrics. The challenge was to fit this friendly, open minded, top-of-the art technical and busy company into an 18th century classified building on Stockholm´s most picturesque walkway by the waterfront.

Chimney工作室是故事的讲述者，借用电影和通讯技术每年为超过60个国家创作6000多部作品。其工作包括创作过程的每一步，从设想到开发、生产，乃至世界一流的后期制作。

设计理念是Chimney家庭 —— 来自世界各地友人的广受欢迎的专业运动礼物和纪念品。办公室内由二手家具和新家具混搭而成，通过低调的颜色和柔软的面料组合在一起。办公室所在的建筑位于斯德哥尔摩滨水区风景如画的走道上，我们面临的挑战是要把这个融洽的、开放的、有着顶级艺术工艺和业务繁忙的公司与这座18世纪的典型建筑融为一体。

Peter Sahlin
Architect at pS Arkitektur, Stockholm, Sweden

彼得·萨林
瑞典斯德哥尔摩pS建筑公司设计师

CONTENTS
目录

▶ Finance 金融

012	SGX Office 新加坡交易所办公室	044	Aserta Aserta办公室
034	Alfa Bank Office 阿尔法银行办公室	050	The Derindere Fleet Leasing The Derindere Fleet金融租赁公司办公室

▶ Information Technology 信息技术

076	Grupa Onet.pl S.A 波兰Onet集团办公室	150	Autodesk Israel-Tel Aviv 欧特克以色列特拉维夫办公室
088	Alert Logic Alert Logic办公室	156	Office Playtech Playtech办公室
102	Onefootball Headquarter 一球网总部	164	Ustream Ustream办公室
110	Yandex Stroganov Yandex斯特罗加诺夫办公室	172	Skype Office in Stockholm Skype斯德哥尔摩办公室
120	Booking.com Office Booking.com办公室	178	Fabrique Delft 代尔夫特Fabrique办公室
130	Google Budapest SPA Office 谷歌布达佩斯温泉办公室	186	Facebook Facebook办公室
138	Cornerstone OnDemand Cornerstone OnDemand办公室	194	Autodesk Shanghai Office 欧特克上海办公室
146	Yudo Office 柳道办公室		

▶ Media 传媒

206	John Brown Cape Town Offices	约翰·布朗媒体公司开普敦办事处
212	Leo Burnett	里奥·贝纳广告公司
218	Chimney Group	Chimney集团办公室
226	FOX International Channels	福克斯国际频道办公室
230	Publicis Groupe	阳狮集团办公室

▶ Design Consulting 设计咨询

240	Office of IND Architects Studio	IND建筑事务所办公室
248	Makhno Workshop	马克诺工作室
258	Arkwright	Arkwright办公室
266	Dizaap Office	Dizaap办公室

▶ Others 其他

276	CP Group	CP集团办公室
284	Met Global	Met全球公司
292	BICOM Communications	BICOM公司办公室
298	Blinq Office	缤客办公室
304	Nexus	Nexus办公室
308	Uralchem Headquarters	Uralchem总部
314	E:MG Office	E:MG办公室

INDEX 索引

Corporate logo image wall
Customizing exclusive corporate culture and temperament

Text Logo
文字型标志

012
110
167

121
206

024
131
220

035
138

077
157
226

企业标志形象墙
——定制企业专属文化气质

051

104

292

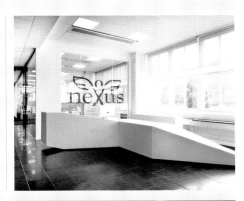
304

194

298

308

277

Graphic Logo
图文结合型标志

285

314

Finance
金融

▲ If you have a dream-creating space
　如果您拥有一个创造梦想的空间

▲ You can decorate it as this
　可以这样装扮它

▲ Or in this way
　也可以这样装扮它

▲ And also in another way
　还可以这样装扮它

▲ Here, you can always find your dreamy office space
　在这里，您总能找到属于自己梦想的办公空间

Finance
金融

PROJECT LOCATION 项目地点	Singapore 新加坡	INTERIOR DESIGN 室内设计	SCA Design (A Member of the ONG&ONG Group) SCA设计（ONG&ONG团队成员）
PHOTOGRAPHER 摄影师	Jiwen Bai 白继文	PROJECT DIRECTOR 项目总监	Brandon Liu 刘俊雄

SGX Office
新加坡交易所办公室

▶ CORPORATE CULTURE 企业文化

The Singapore Stock Exchange (SGX) is widely perceived as a growing gateway to the global market place, and an institution that upholds the strictest standards of transparency. SGX represents the premier access point for managing Asian capital and investment exposure, and is Asia's most international exchange with more than 40% of companies listed on SGX originating outside of Singapore. SGX is led by a team of capable leaders and members who work seamlessly together, to ensure that the right decision is made at the right time.

新加坡交易所（以下简称"新交所"）被广泛认为是一个不断发展的通向全球市场的桥梁，同时是一个有着最严格透明标准的机构。新交所是管理亚洲资本和投资风险的首选平台，也是亚洲最为国际化的交易所，有超过40%的上市公司均来自新加坡以外的地区。新交所拥有才华横溢的领导者和团队成员，他们顺畅开展密切合作，确保在正确的时间作出正确的决定。

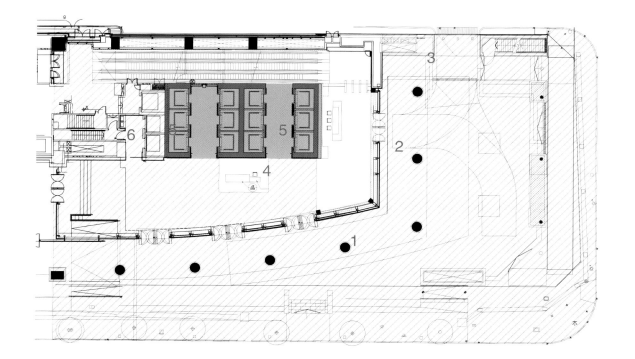

FIRST FLOOR PLAN 一层平面图

1. LCD SCREENS WITH LIGHT BOX
2. REFER TO ARCHITECTURAL DRAWINGS
3. VIP ENTRANCE
4. OPERABLE DURING FIRE EMERGENCY AS SECONDARY FIRE LIFT
5. HANDICAP LOW ZONE LIFT
6. PASSENGER LIFT LOBBY

1. LCD 屏幕含灯箱
2. 参考建筑图纸
3. VIP 入口
4. 火警紧急情况可运行副电梯
5. 障碍低区电梯
6. 客梯大堂

▶ DESIGN CONCEPT 设计理念

When designing the new SGX, insights from staff and customer alike made the design team realize that this traditional institution is undergoing a rapid transformation; this called for entirely new design concepts that facilitated new ways of relationship building, working and approaching technology.

设计新交所新址的时候，员工和客户的深刻见解令设计团队意识到，传统制度正在经受飞速转型；这需要彻底的、全新的设计理念来促成人际创建、工作与接近技术的新路径。

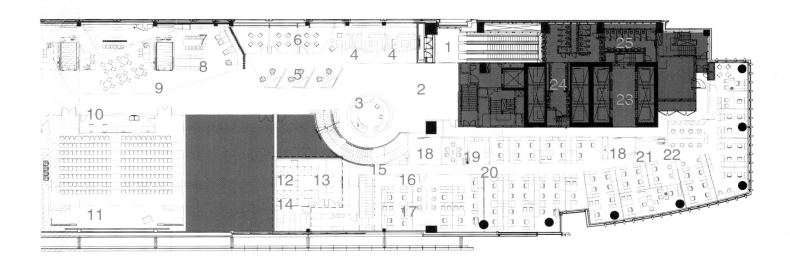

SECOND FLOOR PLAN-1 二层平面图 - 1

- 1. TICKER TAPE
- 2. RECEPTION
- 3. ISLAND RECEPTION
- 4. CLIENT LOUNGE
- 5. OPEN DISC
- 6. MEET OUT
- 7. BAR COUNTER
- 8. COMMUNAL CAFE BENCH
- 9. MARKET PLACE
- 10. PROJECTOR RM
- 11. PROJECTOR SCREEN
- 12. SERVER ROOM
- 13. CRAC ROOM
- 14. UPS ROOM
- 15. CORRIDOR
- 16. LEGAL & COSEC
- 17. GENERAL WORK AREA
- 18. HOLDING AREA
- 19. MINI MARKET
- 20. GENERALBROKER /RM WORK AREA
- 21. PRINT/UTILITY AREA
- 22. GENERAL WORK AREA
- 23. LIFT LOBBY
- 24. MALE TOILET
- 25. FEMALE TOILET

- 1. 行情显示系统
- 2. 接待室
- 3. 接待岛区
- 4. 客户休息室
- 5. 开放区域
- 6. 会议室出口
- 7. 吧台
- 8. 公共咖啡台
- 9. 市场区
- 10. 放映室
- 11. 放映屏幕
- 12. 机房
- 13. 机房空调室
- 14. 供电室
- 15. 走廊
- 16. 法律顾问区
- 17. 综合工作区
- 18. 等候区
- 19. 迷你市场区
- 20. 经纪人工作区
- 21. 打印 / 公共设施区
- 22. 综合工作区
- 23. 电梯大厅
- 24. 男士洗手间
- 25. 女生洗手间

▶ FEATURE 特色

Two working sites were created: a centrally located SGX Centre that focuses on client engagement and a Vista Site that focuses on staff development and team building. The motifs, patterns and textures of the stock exchange display boards were used to create a series of signature motifs to connect the spaces within and across the sites, and capture the importance of communication to the organization.
The SGX Centre also offers a market place, a financial bookshop and refreshment area; not to mention a multipurpose room that functions as office, lounge or dining area.

　　创建了两个工作地点:一个新交所中心，关注客户交流；一个展望中心，关注员工发展与团队建设。该证交所展示板的图案、花样、纹理都被用于打造一系列标志性图案，在地点内外连接各个空间，以此展现沟通与组织的重要性。
　　新交所中心还提供了市场、金融书店和茶点区；还有一个多用途的房间，可作为办公室、休息室或餐厅。

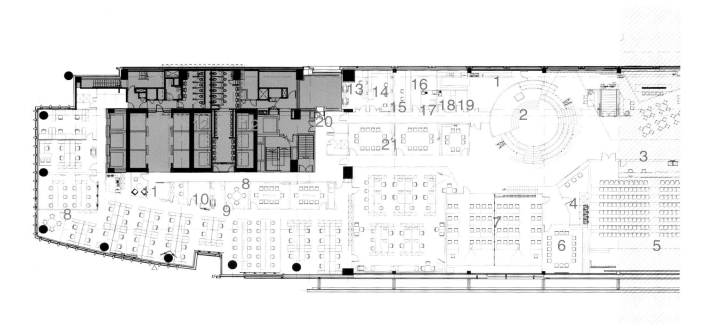

SECOND FLOOR PLAN -2 二层平面图 - 2

1. PANTRY STORE & PREP AREA	8. MEET IN	15. SECURITY RM	1. 餐具室和准备区	8. 会议室入口	15. 安全室
2. IPO STAGE	9. GENERAL WORK AREA	16. FM STORE RM	2. 公开募股展台	9. 综合办公区	16. FM 储藏室
3. PROJECTOR RM	10. PLAYBACK RM	17. NURSING RM	3. 放映室	10. 重放室	17. 护理室
4. STAGE CONTROL RM/AV RACKS	11. HOLDING AREA	18. FIRST AID RM	4. 级控室	11. 等候区	18. 第一急救室
5. EXISTING AUDITORIUM	12. SHAFT	19. PRAYER RM	5. 既有礼堂	12. 通风井	19. 祷告室
6. MEET OUT	13. FM PRINT RM	20. CLEANERS STORE RM	6. 会议室出口	13. FM 打印室	20. 清洁储藏室
7. SGX ACADEMY 2	14. FM MAIL RM	21. MEET OUT	7. 新交所研究院 2	14. FM 收发室	21. 会议室出口

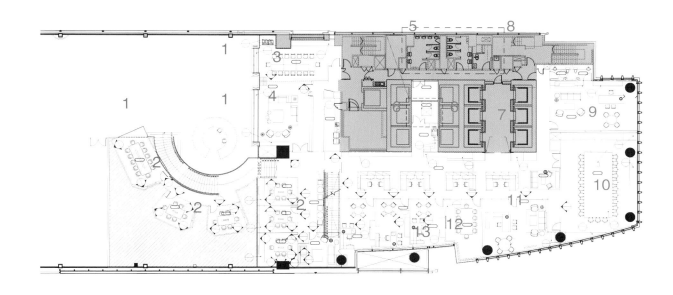

THIRD FLOOR PLAN 三层平面图

1. GLASS WINDOW
2. MEET OUT
3. 16 PAX FORMAL DINING TYPE
4. MULTI PURPOSE ROOM
5. PROVIDE INSULATION UNDER RAISED FLOOR (FOR CONDENSATION)
6. SHAFT
7. FRAME FOR SECURITY PANEL
8. TOILET PACKAGE
9. CHAIRMAN OFFICE
10. BOARDROOM
11. CEO OFFICE
12. MEET IN
13. PRESIDENTS OFFICE

1. 玻璃窗
2. 会议室出口
3. 16座正式餐厅
4. 多功能室
5. 供活地板下隔离（凝结）
6. 通风井
7. 安全嵌板框架
8. 卫浴包装
9. 董事长办公室
10. 董事会议室
11. 首席执行官办公室
12. 会议室入口
13. 总裁办公室

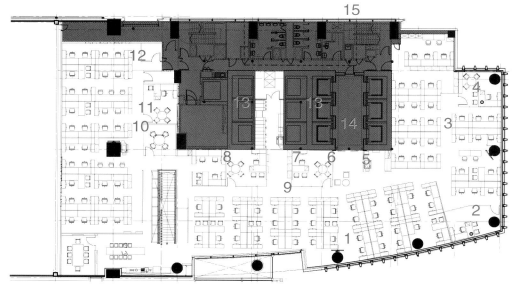

FOURTH FLOOR PLAN 四层平面图

1. POLE MOUNTED
2. SALES TYPE A
3. GENERAL WORKING AREA
4. MARKET DATA
5. PRINT/TILITY AREA
6. LOUNGE AREA
7. SECURITIES TYPE A
8. POST TRADE TYPE A
9. GENERAL WORKING AREA
10. GLASS WHITE BOARD
11. DERIVATIVES TYPE A
12. DERIVATIVES / SALES STORE
13. SHAFT
14. FRAME FOR SECURITY PANEL
15. REFER TO A800 TOILET PACKAGE DRAWING

1. 柱式安装
2. A型销售办公室
3. 综合工作区
4. 市场资料室
5. 打印／公共设施区
6. 休息区
7. A型担保室
8. A型邮递业务室
9. 综合工作区
10. 玻璃白板
11. A型衍生产品室
12. 衍生产品销售店
13. 通风井
14. 安全嵌板框架
15. 参考A800卫浴组装草图

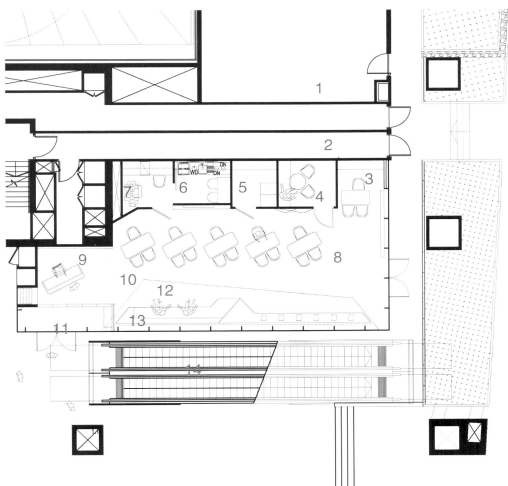

FIRST FLOOR PLAN 一层平面图

1. MANAGEMENT ROOM	1. 管理室
2. INTERNAL EXIT CORRIDOR	2. 内部安全通道
3. BROKER COUNTER	3. 经纪人柜台
4. DISPUTE RM	4. 辩论室
5. EQUIPMENT RM	5. 设备室
6. PANTRY	6. 餐具室
7. MINI OFFICE	7. 迷你办公室
8. COUNTER	8. 柜台
9. SIGNAGE WALL	9. 标识墙体
10. CONCIERGE /INFO COUNTER	10. 礼宾部/信息台
11. ENTRY	11. 入口
12. WAITING AREA /SELF-HELP KIOSK	12. 等候区/自助电话区
13. MOVABLE BENCH SEATING	13. 可移动长椅
14. ESCALATOR	14. 自动扶梯

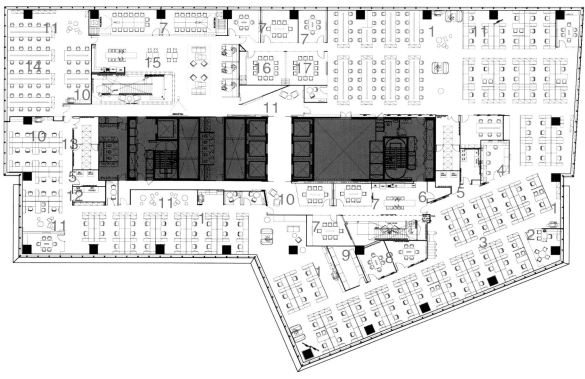

FIFTH FLOOR PLAN 五层平面图

1. PRINT / UTLITY AREA	7. MEET OUT	13. SERVER ROOM	1. 打印/公共设施区	7. 会议室出口	13. 服务器室
2. TECH TYPE A	8. MEET IN	14. PMO HOTDESKING AREA	2. A型技术室	8. 会议室入口	14. 项目管理公用办公区
3. TECH-ESR	9. STORE ROOM TEC(SM)	15. MINI MART	3. ESR技术室	9. 储藏室	15. 迷你市场
4. STAGING AREA	10. HOLDING AREA	16. CRISIS ROOM	4. 集结区	10. 等候区	16. 危机处理室
5. CRAC ROOM	11. OPEN DISCUSSICN AREA	17. SITUATION ROOM	5. 机房空调区	11. 开放讨论区	17. 情景室
6. CLEANER'S STORE	12. DB ROOM		6. 清洁工储物室	12. 配电室	

SIXTH FLOOR PLAN 六层平面图

1. LANDSCAPE	10. MINI MART	19. OPS MAIL ROOM	1. 景观	10. 迷你市场	19. OPS 收发室
2. OPEN ROOF TERRACE	11. CLEANER STORE	20. HOLDIING AREA FOR DELIVERY	2. 露顶露台	11. 清洁储藏室	20. 交货等候区
3. PRINT AREA/UTIUTY AREA	12. NURSING RM	21. PRINT AREA/UTILITY AREA	3. 打印/公共设施区	12. 护理室	21. 打印/公共设施区
4. MEET IN	13. PRINT AREA/UTILITY AREA	22. FM MAIL ROOM	4. 会议室入口	13. 打印/公共设施区	22.FM 收发室
5. STORE ROOM (DEP)	14. STORE ROOM (OC&OPI)	23. FM PRINT AREA	5. 储藏室	14. 储藏室	23.FM 打印区
6. MEET OUT	15. MARKET CONTROL	24. SECURITY	6. 会议室出口	15. 市场管理室	24. 安保室
7. COUNSELING ROOM	16. BOARDROOM	25. FM STORE RM	7. 咨询室	16. 会议室	25.FM 储藏室
8. PRAYER RM	17. PLATFORM	26. LIFT SHAFT	8. 祷告室	17. 月台	26. 升降机井
9. FIRST AID RM	18. MARKET PLACE	27. AHU ROOM	9. 第一急救室	18. 市场	27. 空调机组室

SEVENTH FLOOR PLAN 七层平面图

1. PRINT/UTILITY AREA	7. OPEN DISCUSSION AREA	1. 打印 / 公共设施区	7. 开放讨论区
2. MEET IN	8. STORE RW PRINT/UTILITY AREA	2. 会议室入口	8. 储藏室 / 打印 / 公共设施区
3. MINI MART	9. INTERNAL AUDIT	3. 迷你市场	9. 内部审计区
4. COUNSELING ROOM	10. MARKET SURVEILLANCE	4. 咨询室	10. 市场监管区
5. MEET OUT	11. LIFT SHAFT	5. 会议室出口	11. 升降机井
6. HOLDING AREA	12. AHU ROOM	6. 等候区	12. 空调机组室

▶ FEATURE 特色

Transparency plays an equally important role in the concept design of SGX Vista, with its large, open reception area, glass-walled meeting rooms, and informal mini-markets on each floor. These are designed as informal, collaborative spaces for discussions, meetings, or general congregation areas.
Overcoming the diversity of activities operating within SGX's offices, the new SGX captures the essence of its business while providing a highly efficient and functional workspace for employees.

透明度在新交所展望中心的设计理念中扮演着同样重要的角色，每一层都有大型开放的接待室、带有玻璃幕墙的会议室、非正式迷你市场。这些都设计为讨论、会议、整体集会的非正式协作空间。

为克服由于新交所办公室运作活动的多样性而造成的困难，新的办公室捕捉到其商业精髓，同时为员工提供一个高效、功能性十足的工作环境。

Finance
金融

PROJECT LOCATION 项目地点	Moscow, Russia 俄罗斯莫斯科	DESIGN COMPANY 设计公司	IND Architects IND建筑事务所
PROJECT AREA 项目面积	2630 m²	OFFICE CONSTRUCTION 建筑公司	RD Construction RD建筑

Alfa Bank Office
阿尔法银行办公室

▶ CORPORATE CULTURE 企业文化

Alfa Laboratory is a special unit of Alfa Bank engaged in electronic business. Alfa Laboratory creates innovations for the Alfa Bank, so employees are creative, educated, active mentally and physically. Everyone can make a proposal, if you have a good idea you can share it with the bosses—there is not that heavy job hierarchy like financial organizations commonly have.

阿尔法研究室是阿尔法银行下属的一个特殊单位，从事电子商务。阿尔法研究室为阿尔法银行提供灵感，所以员工们都非常有创造力、受过良好教育且身心活跃。每个人都可以提出建议，如果你有好主意，可以与老板们分享——这里并不像一般的金融机构那样有严格的等级制度。

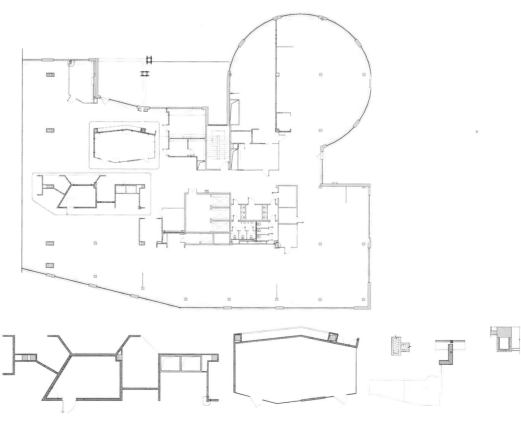

▶ DESIGN CONCEPT 设计理念

The concept of the interior is based on superheroes and street art—the components which Laboratory employees can relate themselves to. The finishing materials used are as follows: textured concrete, wood, perforated metal, various kinds of glass—clear, dim, and patterned. A bright carpet tile facilitates the navigation—meeting zones are colored, while various circles in an open space zone help employees to find required groups and departments.

　　室内设计理念基于超级英雄和街头艺术，研究室的员工可以将这些成分与自身联系起来。运用到的材料如下：有纹理的混凝土、木材、穿孔金属板、不同种类的玻璃——清晰的、模糊的、有花纹的。明亮的地毯瓷砖便于人们找到正确的方向——会议区被色彩装点，开放空间的不同圆圈帮助员工们找到所需的组织和部门。

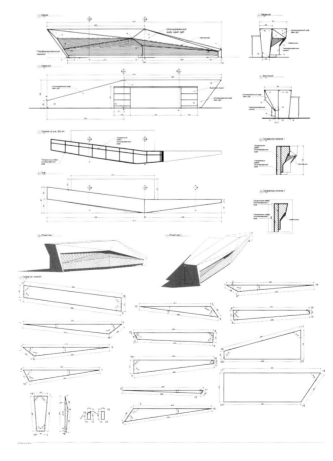

▶ FEATURE 特色

One of the unique features of the office designed by architects of IND Architects for one of Alfa Bank's branches is, first of all, its creativity which is incredible for the financial sector. Functional, striking, and innovative, it will be a place to work in for young and vigorous employees of Alfa Laboratory.

Walls of the office combine several functions: 1) a decorative function—bright wall murals featuring superheroes and comic books; 2) a practical function—a special surface where you can write with felt-tip pens; a bulletin board material has been pasted on some walls, where employees can fix their materials; and 3) an informative and motivational function—a wall with quotes of great people.

  由IND建筑事务所设计的阿尔法银行分支办公室的特色，首先是其创造性，这对于金融部门来讲是不可思议的。功能十足、醒目、创新，这将成为阿尔法研究室年轻、有活力的员工们工作的地方。

  办公室的墙壁具备多重功能：1) 装饰功能——上面有着超级英雄和漫画图案的明亮壁画；2) 实用功能——可以用记号笔在特殊表层上书写；一些墙上贴着布告栏，员工可以在上面整理材料；3)教育、激励功能——墙上引用了伟人、名人的格言。

A distinguishing feature of the Laboratory is that employees can not only work on their work places, but take an active part in brainstorms and meetings as well. This feature has identified the laying out of the office—there are many meeting zones and buzz session zones; the game zone to have a rest with a ping-pong table, various board games, and carpet-covered walls to play darts; and two outdoor porches with gleamy furniture here. There is an open meeting hall in the center of the office. There are LCD screens in front of meeting rooms which display if the rooms are available or not.

研究室的一大显著特色是员工不仅可以在工作场所办公，也可以积极参加头脑风暴和会议。这个特色确定了该办公室的布局——许多会议空间及小组漫谈会；游戏区有一张乒乓球台、众多桌游、可以射飞镖的附有毯子的墙壁；两个摆放着金光闪闪家具的户外走廊。办公室中央有一间开放式会议大厅。会议室前方都有液晶屏幕，显示该房间是否可用。

Various lighting solutions have been implemented in the office—linear light in the open space zone, LED backlighting in a hallway, and soffits in the game and the presentation zones. The classy, dynamic, and functional design with striking elements and interesting details—this is not merely an office, but a really comfortable place to create unusual and contemporary solutions too.

办公室应用了各种各样的照明方案——开放空间区域的线性光、走廊处的LED背光、游戏区及展示区的底光。优雅、动态、功能型的设计拥有引人注目的元素及趣味性十足的细节——这不仅仅是一间办公室，还是一个可以打造不同寻常当代方案的真正舒适的地方。

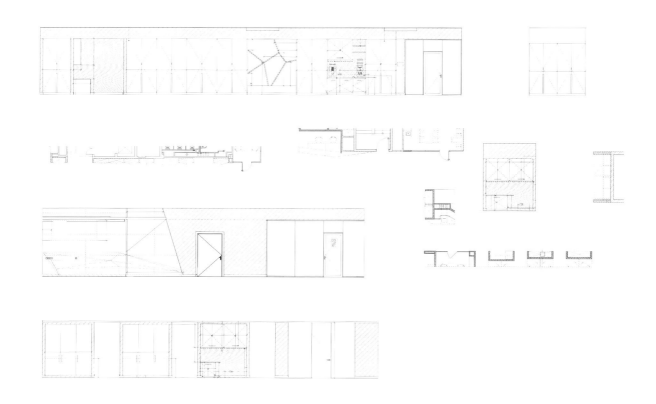

FLOOR PLAN 平面图

1. OPEN SPACE
2. COFFEE POINT
3. WARDBROBE
4. PRINT ZONE
5. MEETING ROOM
6. KITCHEN
7. RELAX ROOM
8. RECEPTION
9. CABINET
10. VIP RECEPTION
11. SERVER
12. WC

1. 开放空间
2. 咖啡点
3. 衣帽间
4. 打印区
5. 会议室
6. 厨房
7. 休息室
8. 接待室
9. 橱柜
10. VIP 接待室
11. 服务室
12. 洗手间

PROJECT LOCATION 项目地点	Mexico 墨西哥	DESIGNERS 设计师	Juan Carlos Baumgartner, Shantal Bravo Vega 胡安·卡洛斯·鲍姆加特纳、杉塔·布拉沃·维加	DESIGN COMPANY 设计公司	Space Space设计公司	FURNISHING 家具设计	Herman Miller, Steelcase 赫曼·米勒公司、Steelcase公司
PROJECT AREA 项目面积	3892 m²	PHOTOGRAPHER 摄影师	Paul Czitrom 保罗·奇特罗姆	LIGHTING 照明设计	Luz en Arquitectura 卢资·恩建筑公司	CONSTRUCTOR 建造师	Alpha Hardin 阿尔法·哈丁

Aserta
Aserta办公室

▶ CORPORATE CULTURE 企业文化

The Financial Group Aserta emerges from the merger of Afianzadora Aserta and Afianzadora Insurgentes forming one of the largest Surety Groups in Mexico, which required remodeling their offices and designers took advantage of the change to strengthen the company culture and create an innovative and transparent form of improving team work and the productivity of the users.

Aserta金融集团由Afianzadora Aserta和Afianzadora Insurgentes两家公司合并而成，是墨西哥最大的担保集团之一，他们需要重塑办公室，设计师利用这个机会强化其企业文化并且打造一种有助于提高团队合作及使用者生产率的创新通透形式。

▶ DESIGN CONCEPT 设计理念

The design starts from the 8th floor which is a central floor and functions as a working lounge and the heart and soul of the project; on this floor not only is it possible to promote wellbeing with balanced meals promoting the health of the user, but there are also new forms of interacting with others, where hierarchical barriers are broken down. From this floor it was possible to spread this communication and transparency to the 7th and 9th floors, through a focal point staircase of white marble connecting the 3 floors, also integrating a large window that lets in lots of light and offers a great south facing view of Mexico City.

该设计从第八层开始，位于中心楼层，起到工作休息室的用途，是整个项目的核心与灵魂；在该层，不仅可以依靠平衡膳食来促进使用者的健康，提升幸福感，而且还有与他人对话的新方式，分层障碍被解除。从这层起，这种交流和穿透性通过连接三个楼层的白色大理石楼梯延伸到了七层和九层，该楼梯同时让阳光从一面大窗户射入，形成了面朝墨西哥城的南向风光。

▶ FEATURE 特色

It is very important to permeate the culture and they were collaborated with in order to produce icons of the intellectual, artistic and musical culture of the world that would serve as inspiration for innovation in the work processes. This was done through a treatment of stickers on glass and on walls, with music icons on the 7th floor, visual artists and painters for the 8th floor and for the 9th floor people who have caused change or been turning points because of their inventions or discoveries. The image in general creates an environment full of life and color that in the design of both carpets and ceilings sends the message that it is an innovative and visionary company that through this project reaffirmed the company culture and improved its productivity and the communication among users.

将文化渗入其中非常重要，他们相互合作，创作世界上智慧、艺术及音乐文化的图标，为工作过程中的创新提供灵感。七层的音乐图标以及八层、九层那些因其发明或发现而对世界产生改变或转折点的视觉艺术家及画家的肖像通过玻璃及墙壁上的贴纸处理予以实现。整体意象打造了一处充满生机与色彩的环境，地毯和天花上的设计传递了一种讯息，这是一家创新、有远见的公司，通过这个方案，再次确认了该公司的企业文化、提升了生产效率以及与使用者之间的交流。

Finance 金融

PROJECT LOCATION 项目地点	Istanbul, Turkey 土耳其伊斯坦布尔	DESIGN TEAM 设计团队	Serter Karataban, Ceyhun Akın, Murat Özbay, Ceyda Atıcı 赛特尔·卡拉塔班、塞汉·阿克、缪拉·奥贝、慧达·阿提斯		
PROJECT AREA 项目面积	2150 m²	PHOTOGRAPHER 摄影师	Mehmet Ince 穆罕默德·恩斯	DESIGN COMPANY 设计公司	TeamFores TeamFores设计公司

The Derindere Fleet Leasing
The Derindere Fleet金融租赁公司办公室

▶ CORPORATE CULTURE 企业文化

The DRD fleet leasing company was founded in 1998 and in 15 years it has become Turkey's largest locally financed leasing company, with a fleet of some 26.000 vehicles. Their present office having trouble coping with the company's growth, plans were made at the beginning of the year 2013 to move it to another building situated on a 2.500 square meter plot.
The visionary senior management of the Derindere Company, who has already contributed many innovative ideas and practices to their sector have contacted TeamFores for the design of a flexible and modern office, which could face the mid-term growth objectives of the company, while boosting personnel productivity to the highest level.

The DRD Fleet 金融租赁公司成立于1998年，15年时间，它一跃成为土耳其当地最大的金融租赁公司，拥有大约26000辆车。他们现有的办公室未能跟上公司的发展步伐，2013年初便确定计划将其迁至另一栋图测面积2500m²的大楼。

Derindere公司的高级管理人员非常有远见，为他们部门贡献了很多创新的灵感和实践，他联系TeamFores公司设计一个灵活且现代化的办公室，吻合公司中期的成长目标，同时促使员工生产效率翻升至最高水平。

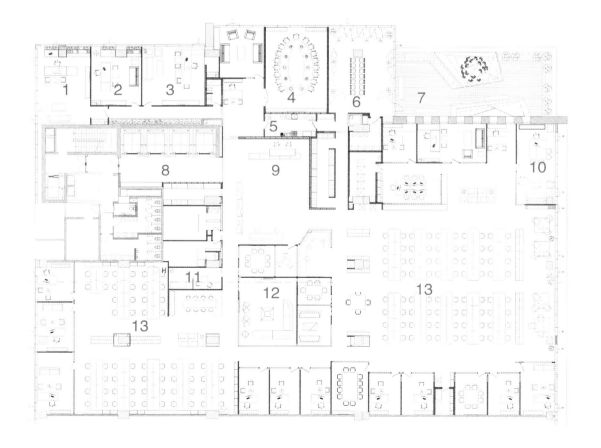

FLOOR PLAN 平面图

1. PRESIDENT OFFICE — 1. 董事长办公室
2. V. PRESIDENT OFFICE — 2. 副董事长办公室
3. CEO OFFICE — 3. 执行总裁办公室
4. BOARDROOM — 4. 会议室
5. KITCHEN — 5. 厨房
6. CAFETERIA — 6. 自助餐厅
7. TERRACE — 7. 阳台
8. ENTRANCE — 8. 入口
9. RECEPTION — 9. 前台
10. GM OFFICE — 10. 总经理办公室
11. DOCTOR — 11. 医务室
12. MEETING LOUNGE — 12. 会议休息室
13. OPEN OFFICE — 13. 开放式办公室

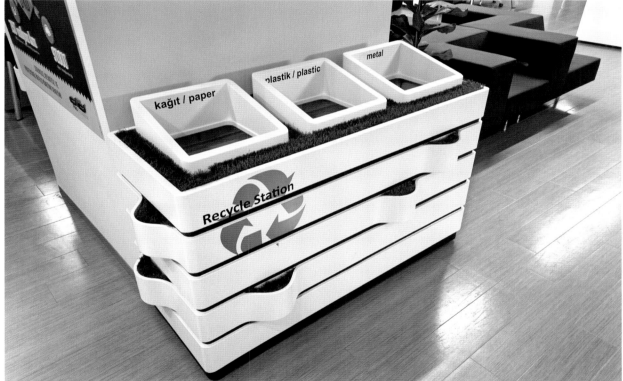

▶ DESIGN CONCEPT 设计理念

The main factor that makes up the specificity of the project consists of providing more space for social and smaller areas for working activities, thus contributing to top level work motivation and increased productivity. The design takes account of the changing working habits of today's employees, providing clever and simple solutions to this new situation. The rooms are conceived as multi-functional units so as to answer to as many needs as possible, thus diminishing the space needed for a project of this size while leaving the possibility to open up spaces to new potential functions. The success of the design of the social premises can be seen in the increased productivity resulting from the bringing together of people who would not usually know each other despite the fact that they work in the same office and the fruitful interaction thus created.

构成该项目特异性的主要因素包括提供更多的社交空间及更小的工作活动区域，如此便有助于增加工作积极性、提高生产率。考虑到当今时代员工们不断变化的工作习惯，该设计提供了巧妙而又简单的方法。为符合尽可能多的需求，这些房间被构想为多功能的小单元，这样就减少了这种规模项目所需要的空间，同时保留开发具有新潜在功能空间的可能性。尽管人们在同一间办公室工作，但他们相互间并不太了解，将其聚在一起，由此，社交场所设计的成功在不断提高的生产率上便可见一斑，同时打造了多产的互动。

▶ FEATURE 特色

The office takes maximum advantage of cutting edge technology, with automation systems monitoring mechanical conditions and lighting according to the number of users and users scenarios so as to minimize unnecessary energy consumption. While avoiding needless use of equipment, this well balanced use of energy also results in increased acoustic comfort.

All furniture used in the project was designed so as to answer present and foreseeable needs, taking account of the Derindere Company's 16 years of experience as well as its growth potential. The furniture was designed and created without overlooking the project's educational and social responsibility dimension. It relieves users from being bound to a

　　该办公室最大限度地利用了切边工艺，根据用户人数及场景，以自动化系统监控机械状态和灯光设备，以便将不必要的能量损耗最小化。为了避免设备不必要的使用，能量的均衡运用提高了听觉舒适度。

　　考虑到Derindere公司16年的资历及其增长潜力，该项目中的所有家具都是定制的,以便符合现在及将来的需求。家具的设计和打造并没有顾及该项目的教育及社交责任。这样便减少了使用者在一个既定房间或办公桌的束缚感，打造了一个自由、人性化的办公空间和环境。这些家具的运用会让使用者在里面走来走去，避免了日常固定不动的工作环境。从办公桌到

given room or desk, creating a free and holistic office space and environment. The use of the furniture will lead the user to move around, thus avoiding a routine and stationary working environment. From working desks to meeting lounges, printer units to moveable caissons, waste recycling units to file storage systems, everything was specifically designed and created for this project.

Another element that can be considered as a contribution to office social life consists of the prayer rooms. Prayer rooms have been conceived as meditation rooms open to people of all religions. Many of Turkey's most important artists and designers have undeniably contributed to the project, under the leadership of TeamFores. Each room is decked with carefully chosen works from Turkey's most renowned painters, which practically turns the office into an art gallery.

会议休息室，从打印机装置到可移动沉箱，从废物回收设置到文件存储系统，一切都是为这个项目特别设计打造。

可以看作办公室社交生活另外一个贡献元素的是礼拜室。礼拜室对所有有信仰的人开放，作为冥想室之用。在TeamFores公司的带领下，土耳其众多重量级艺术家和设计师毋庸置疑地为该项目做出了贡献。每个房间都装点着土耳其著名画家精选的画作，几乎将这个办公室变成了画廊。

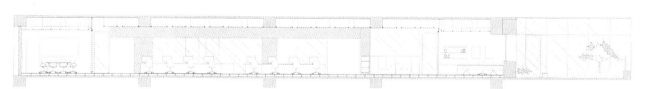

SECTION AA 立面图 AA

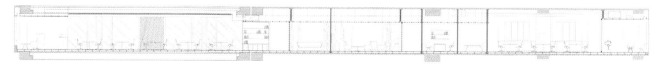

SECTION BB 立面图 BB

Information Technology
信息技术

▲ If you have a dream-creating space
　如果您拥有一个创造梦想的空间

▲ You can decorate it as this
　可以这样装扮它

▲ Or in this way
　也可以这样装扮它

▲ And also in another way
　还可以这样装扮它

▲ Here, you can always find your dreamy office space
　在这里，您总能找到属于自己梦想的办公空间

Information Technology
信息技术

PROJECT LOCATION 项目地点	Poland, Kraków 波兰克拉科夫	DESIGN COMPANY 设计公司	Mocolocco Mocolocco公司	DESIGN TEAM 设计团队	Karolina Tkocz, Karolina Kokot, Anna Korzeniowska, Marcin Rzycki 卡罗丽娜·托卡兹、卡罗丽娜·科科特、安娜·克泽尼欧卡、马尔钦·里慕
PROJECT AREA 项目面积	4,500 m²	PHOTOGRAPHER 摄影师	Igor Stanisławski 伊戈尔·斯坦森塔斯科		

Grupa Onet.pl S.A
波兰Onet集团办公室

▶ CORPORATE CULTURE 企业文化

Onet.pl Group, founded in 1996 is a leading polish online publisher focusing on information, entertainment, communication, mobile and online advertising. Onet is the biggest polish medium online which offers a lot of developed thematic services. Most of them are the market leaders in their categories. Onet reaches over 13 million unique users per month, which gives nearly 67 percent of all internet users in Poland. Onet pages gain nearly 2,4 billion page view. For this reason, Onet group offer advertisers the biggest and most attractive premium advertising space in polish network. Onet.pl is one of the strongest and most recognizable polish media brands.

波兰Onet集团，成立于1996年，是波兰在线出版商中的领头者，专注于信息、娱乐、通信、移动和在线广告。Onet集团是波兰最大的在线媒介，提供很多成熟的主题服务。他们大部分是所属行业中的市场领导者。目前Onet集团每月拥有独立用户达1300万以上，占据了波兰互联网用户近百分之67%。Onet页面获得近2.4亿页视图。出于这个原因，目前Onet集团为广告

主提供了波兰网络上最大和最有吸引力的高端广告空间。波兰Onet集团是波兰媒体中最强且最知名的品牌媒介。

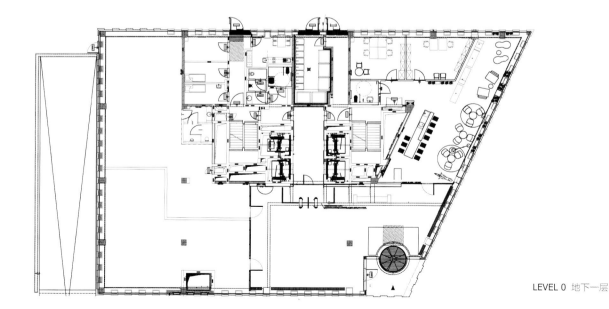

LEVEL 0 地下一层

▶ DESIGN CONCEPT 设计理念

It is no secret that the portal onet. pl aspires to rank on google. Google Office recognizable are practically all over the world, are famous for their informal space. Investor wanted his office interiors were also, unusual, energizing, remaining in memory , associated with the portal.

The task of the designer was to design and connect over the informal, which have boosted the creativity of employees and give them moments of relaxation with spaces to work in the intenty.

This task was difficult because the Office onet. pl employ 650 workers on the surface of a 4500m^2. Functionally arrange space for conversations, meetings and work to come up with ideas and the thought of guiding.

The inspiration to become the elements that surround us and at the same time are characterized by individual divisions of the company. Uniqueness, lightness and panache creating our own through the use of mainly non-finishing materials, furniture (mainly Polish designers) not forgetting to design space has always been a functional and ergonomic.

波兰Onet想要在谷歌上得到排名这并不是秘密。已知的谷歌办公室几乎遍布全球，以其非正式空间著称。投资者希望他办公室的内部也一样与众不同、活力充沛、令人印象深刻，与其门户网站相关。

设计师的任务是设计并进行非正式联结，提高员工的创造性，给他们休息的空间，让他们心无旁骛地工作。

这项任务很艰巨，因为Onet波兰办公室的面积为4500m^2，拥有650名员工。从功能上安排对话、会议、提出工作建议及指导思想的各个空间。

灵感成为我们周边的元素，公司不同部门各有特色。通过使用非精工材料与家具（多为波兰设计师设计），以其独特、明亮、灿烂打造了一个符合功能需求与人体工学的空间。

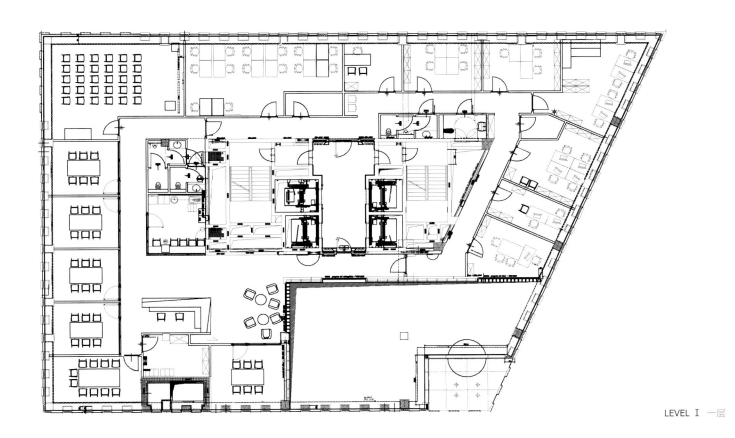

LEVEL I 一层

▶ FEATURE 特色

First floor-colors black and yellow logo as a background color of white.
On each of these floors materials and their structure and colors are appropriate for the element.
On most floor ceilings are open. This solution has helped get more space and loft offices in nature where the majority of working young people.
On the ground floor of an office building is a common space for employees Playroom.
A place where you can enjoy a cup of good coffee, sit down in a comfortable chair, put on the picture taken in and play Pinball machines but also in this field, you can arrange a meeting.
Because if the worker to perform his duties, must sit behind the desk? Whether ridding it of any distractions or conditions to rest will make will work more efficiently?The designer doesn't think so. Experience shows something quite the reverse. The longer you don't come away from the computer (often just there is too much where . . .), the harder it is for us to focus on. The more similar to each other days spent at work, the less enthusiasm and desire to perform their duties.

一楼——黑色和黄色商标，白色为底。

每个楼层的材料及其结构、颜色都与元素匹配得宜。

大多数楼层的天花板都是开放式的。这个方案可以获得更多的空间，以及大多数年轻人乐于工作的自然环境。

办公室一楼是一个供所有员工出入的娱乐室。

在这里，你可以享用一杯上好的咖啡，在一把舒适的椅子上安坐，放上刚刚拍摄的照片，玩弹球机，还可以安排一场会议。

员工履行自己的职责一定要坐在办公桌后面吗？摆脱所有干扰去休息是不是可以让工作更有效率？设计师并不这么认为。经验恰恰表明不是。你紧盯着电脑不离开的时间越长，越不容易集中注意力。每天工作的情景越相似，他们履行职责的热情和渴望越少。

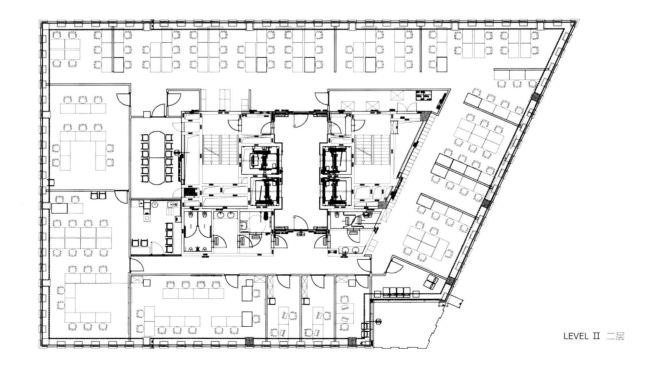

LEVEL Ⅱ 二层

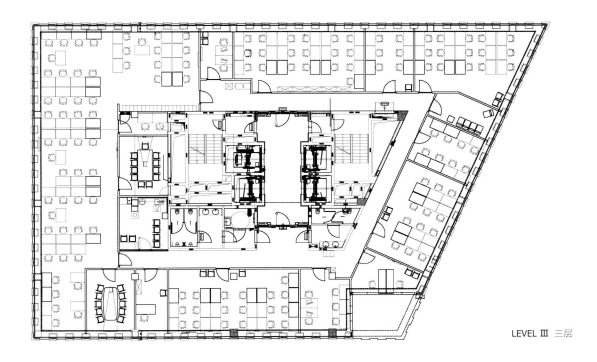

LEVEL III 三层

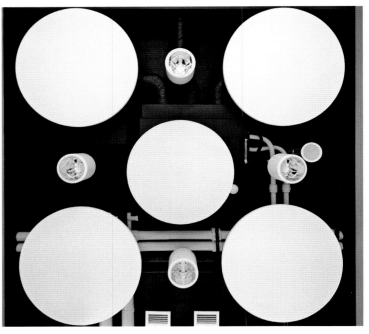
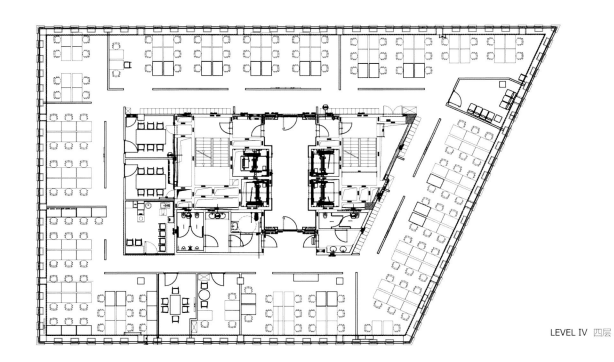

LEVEL IV 四层

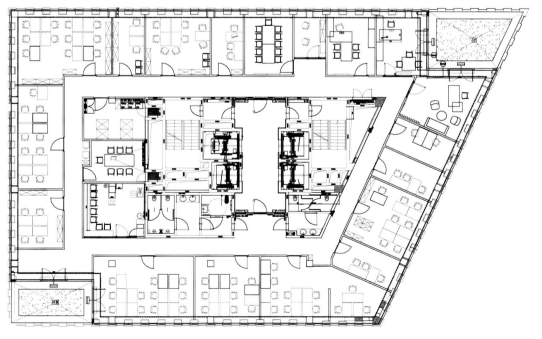

LEVEL V 五层

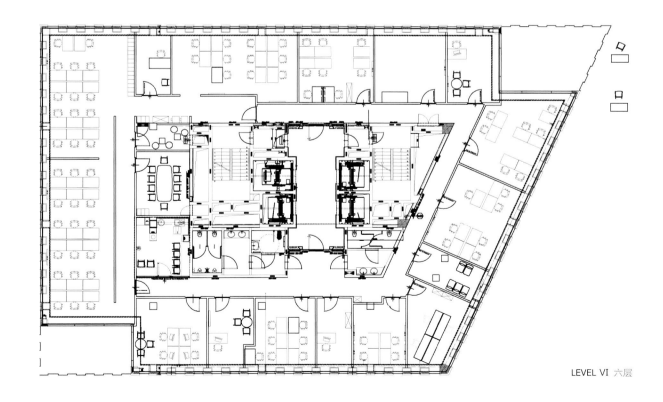

LEVEL VI 六层

Information Technology
信息技术

PROJECT LOCATION 项目地点	Cardiff, UK 英国加的夫	DESIGN COMPANY 设计公司	Space & Solutions Space & Solutions公司
PROJECT AREA 项目面积	1000 m²	PHOTOGRAPHER 摄影师	The Electric Eye The Electric Eye摄影

Alert Logic
Alert Logic办公室

▶ CORPORATE CULTURE 企业文化

Alert Logic has more than a decade of experience pioneering and refining cloud solutions that are secure, flexible and designed to work with hosting and cloud service providers. We deliver a complete solution that lives in the cloud, but is rooted in real expertise. They're a vibrant, fast growing and successful IT Security company and see themselves as a next generation security company' and wanted their new workspace to reflect this.

  Alert Logic有超过十年先进精练的云处理方案，安全、灵活，与主机和云服务供应商一起共事。我们在真实的专业技术中，提供全面的云端解决方案。他们是充满活力、快速增长且成功的互联网安全公司，视其自身为下一代领导者，并希望他们的工作空间能反映出这一点。

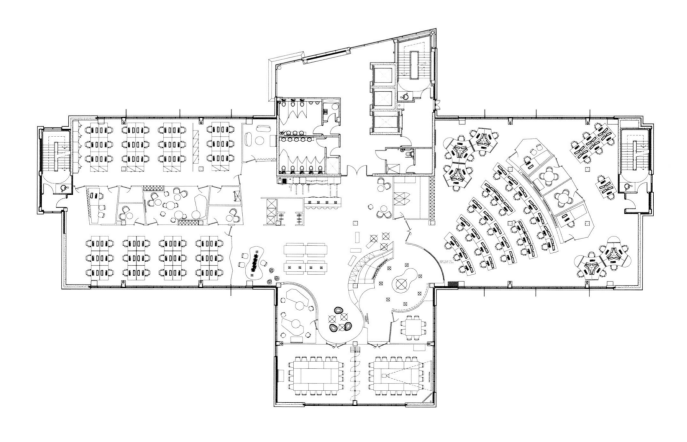

▶ DESIGN CONCEPT 设计理念

In May 2014 Alert Logic invited Space & Solutions to partner with them to create their Cardiff based EMEA Head Office. This European Operations Centre which was a new venture within the UK was set up to operate on a follow-the-sun basis with US operations. An innovative design arrangement was introduced providing a "Journey through Alert Logics workspace".

The journey, which started in a coffee shop style front of house, was designed to enhance the sales experience for clients and to encourage potential staff to join the team, in what is a highly competitive market. It follows the company's history, emphasizing significant stages of company growth and achievements, embedding the culture of the company within this new space.

2014年5月Alert Logic邀请Space & Solutions作为合作伙伴，共同完成他们在英国加的夫市的欧洲总部办公室。在英国刚刚成立的欧洲运营中心准备像美国运营中心一样运作。引进创意设计布局，打造一个"Alert Logics办公室之旅"。

这段旅程开始于咖啡店风格的房子前面，旨在于一个高度竞争的市场中提升客户销售经验、鼓励有潜质的员工加入团队。遵循公司的历史，强调公司成长进步的重大阶段，在这个崭新的空间植入企业文化。

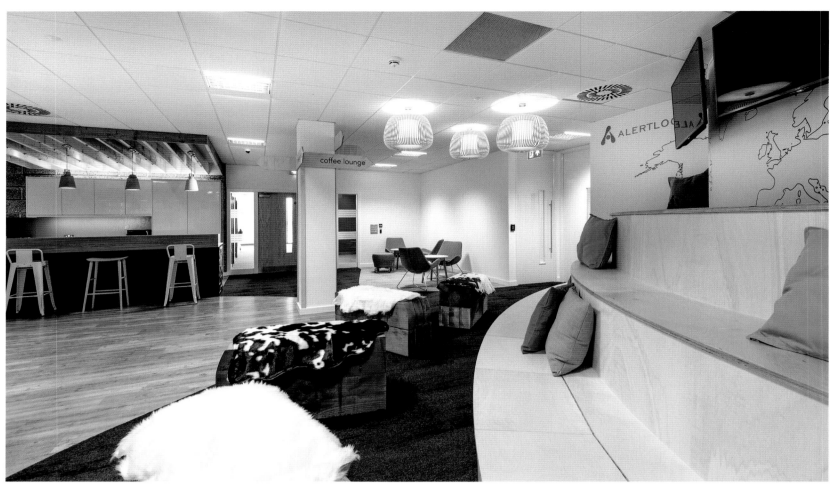

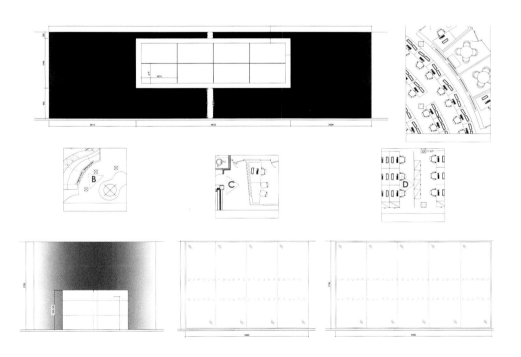

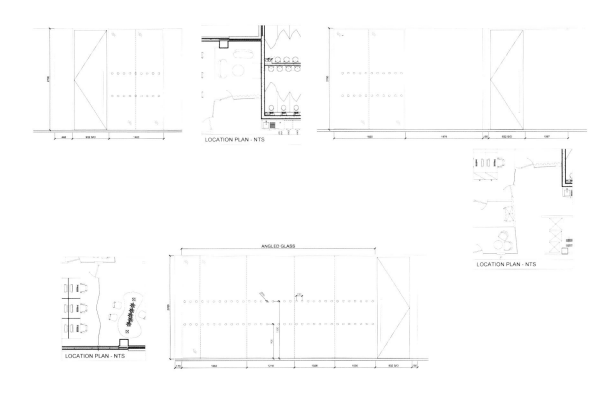

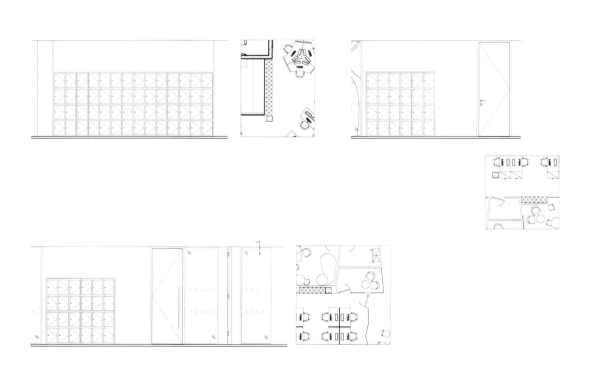

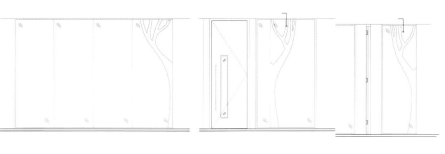

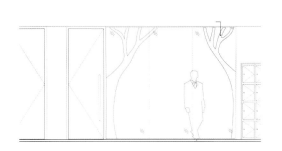
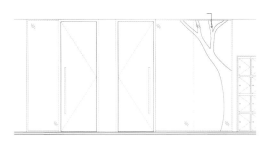

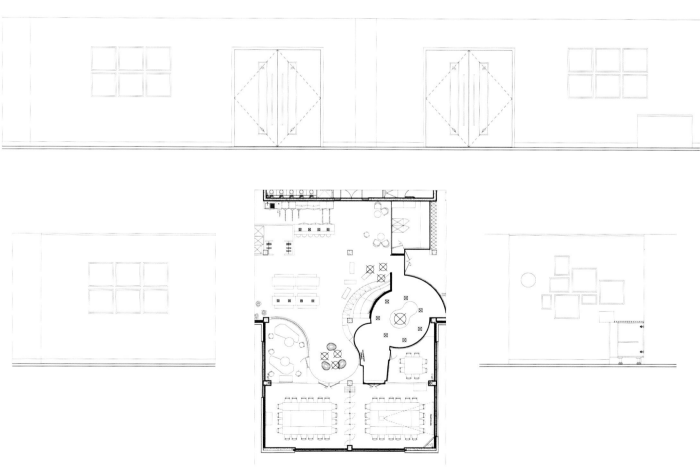

▶ FEATURE 特色

Sarah O'Callaghan, lead designer and Managing Partner of Space & Solutions said "In a world where so much is expected we always endeavour to create professional & exciting workspaces which stand out from the crowd. Alert Logic needed to grow their business in the UK, attracting new clients who require the confidence that a long standing company offers."

Designed to be the new home for approximately 100 staff with facilities to accommodate an increase in staff headcount as the business grew was an important factor. The space created was vibrant and engaging promoting collaboration amongst all the teams.

The entire workspace is free-flowing to unite the workforce and provide a professional environment with an abundance of flamboyance.

The colors and materials used were introduced to unite the Texan and Welsh heritage. One of the meeting rooms, rather unconventionally fitted out with an unusual array of soft furnishings, was named "The Cwtch".

Sarah concluded "Alert Logic embraced everything we set out to do and worked with us to create a workspace which is an incredible place to work. We are proud to have designed and built a unique workspace where the staff at Alert Logic can thrive and truly love their workspace."

Space & Solutions 首席设计师兼经营合伙人Sarah O'Callaghan说："在这个期待过多的世界，我们一直努力创建专业、振奋人心且脱颖而出的工作空间。Alert Logic需要在英国开展业务，吸引需要公司悠久历史建立信心的新客户。"

为大约100名员工设计这个设施齐备的新房子以适应业务增长带来的人员增加是一个重要因素。打造的空间充满生机，魅力动人，促进所有团队间的通力合作。

整个工作空间自由通畅，将所有员工团结在一起，提供了一个华丽丰富的专业环境。

为结合德克萨斯和威尔士的传统，引入了一些色彩和材料。其中一间会议室不依惯例地配备了一组不寻常的软装饰，命名为"The Cwtch"。

Sarah的结论是"Alert Logic接受我们所做的一切，和我们一起工作，创建一个不可思议的办公空间。我们很荣幸设计并建造了这样一个独特的办公空间，这里的员工可以在此茁壮成长并真正热爱他们的办公空间。"

Information Technology 信息技术

PROJECT LOCATION 项目地点	Berlin, Germany 德国柏林	DESIGNERS 设计师	Tanjo Kloepper, Eugenia Zimmermann 塔尼奥·克洛珀、尤金妮亚·齐默尔曼	DESIGN COMPANY 设计公司	TKEZ architecture & design TKEZ建筑设计公司
PROJECT AREA 项目面积	1393.5 m²	PHOTOGRAPHER 摄影师	Benjamin A. Monn 本杰明·阿·莫恩		

Onefootball Headquarter
一球网总部

▶ CORPORATE CULTURE 企业文化

Munich based Architects TKEZ architecture & design have designed and built a new Headquarter for the world's leading football community Onefootball. The Onefootball App connects more than 14 million football fans in over 200 countries and allows them to follow their favorite team—anytime and anywhere. TKEZ's task was to create a stimulating and at the same time professional working environment for 90 employees on a surface area of 15,000 square feet.

总部位于慕尼黑的TKEZ建筑设计公司，设计并建造了世界领先的足球联盟一球网的新总部。一球网的应用程序将200多个国家1400多万球迷联系在一起，让他们随时随地追随自己喜爱的球队。TKEZ公司的任务是在15000平方英尺（1393.5m²）的面积内为90名员工打造一个既令人兴奋又具有专业性的工作环境。

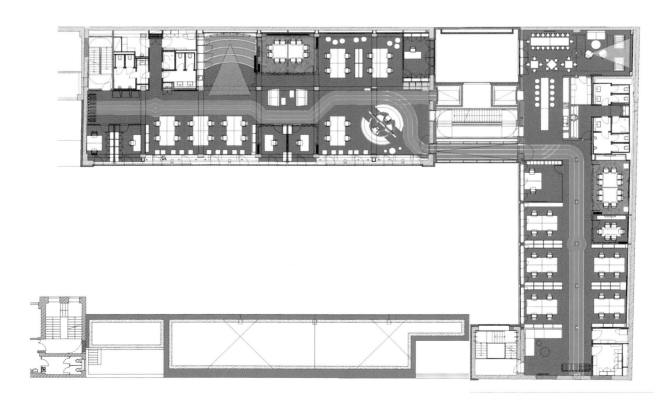

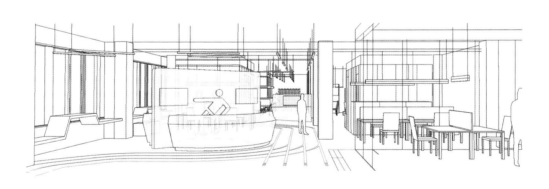

TOUCH DOWN	OPEN OFFICE	RUNWAY	TEA KITCHEN	ARENA
即用工作区	开放式办公室	跑道	厨房	竞技场

▶ DESIGN CONCEPT 设计理念

Communication is creation
The office was designed to be a light and open, multi-functional work environment which enhances communication and teamwork among the coworkers and represents the young and enthusiastic spirit of the company.

交流即是创造
这间办公室旨在设计成为一个采光充足、开放、多功能的工作环境，从而提升同事间的交流与团队合作，同时展现出公司年轻且充满激情的精神。

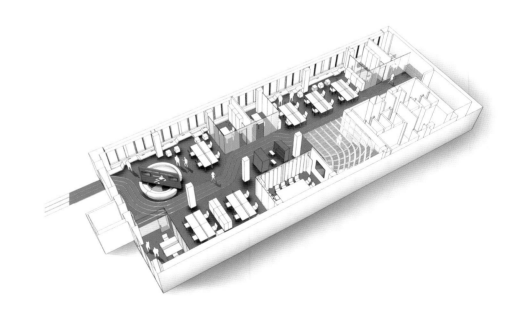

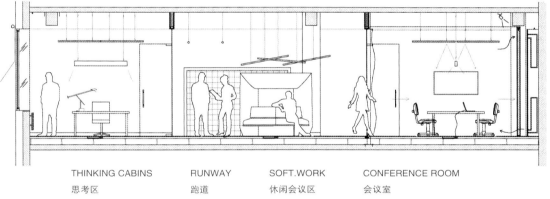

| THINKING CABINS | RUNWAY | SOFT.WORK | CONFERENCE ROOM |
| 思考区 | 跑道 | 休闲会议区 | 会议室 |

▶ FEATURE 特色

A dynamic grass green runway is curving through the generous open floor plan connecting the office's various working zones and areas of relaxation. The ample space is rhythmized by transparent floor to ceiling glazed single and double unit thinking spaces, offices and meeting rooms which are carefully placed in order to fluently create a subtle zoning for the diverse teams. If a more private and calm work environment is required these areas can be closed by translucent curtains. Artificial grass floors generate a special reference to the companies theme: Football.

Long linear Touch Down areas of varying height along the façades offer the employees a change from their assigned desks to a location where they can have spontaneous meetings with their coworkers or simply work towards the light filled green courtyard.

Sofa areas at the back end of the offices offer informal seating to meet and discuss in a casual surrounding. The walls in this areas have special magnetic and rewritable surfaces to support brainstorm sessions and the generation of ideas and visions which can be sketched right on the walls to be communicated to the team mates.

一条充满活力的草绿色跑道曲曲折折地穿越宽阔、开放的平面图，连接办公室各个不同的工作区与休息区。透明地板及镶有玻璃的单人或双人思考空间、办公室、会议室等让这个宽敞的空间律动十足，这些区域经过精心设置，以便为不同团队顺利打造一个微妙地带。如果需要更私密、安静的工作环境，可以通过半透明的窗帘来封闭这些区域。人造草坪地板则形成了一种特殊方式，关联了公司的主题：足球。

沿立面不同高度的长线型落地区为员工提供了与指定办公桌不同的地方，在那里他们能够和同事自如地会谈或者只是朝向阳光充足的绿色庭院工作。

办公室后方的沙发区域让人们可以在一个非正式的休闲环境里会面、讨论。这些区域的墙壁带有强力磁性，并有着可擦写的表层，人们可以通过浏览墙壁上的东西和团队成员交流，促使头脑风暴会议、灵感与愿景的产生。

Onefootball Arena

The heart of the new HQ is the Onefootball Arena. It is a multi functional area which serves as the main space for presentations, meetings, interactive work sessions, football events and company gatherings. The programmer or designer teams gather here and have workshops on the large 4x4m screen on which the latest ideas and work can be projected. At football games the arena with it's colored backlit roof and surround sound system fills with employees, their families and friends and creates a very special and vibrant environment.

一球网竞技场

一球网新总部的核心是它的竞技场。这是一个集展示、会议、互动式工作对话、足球赛事及公司聚会等多功能为一体的区域。程序员和设计师团队聚在这里,朝着4m x 4m的屏幕进行研讨,那上面投射着最新创意及工作内容。进行足球比赛的时候,这个带有彩色背光屋顶、环绕立体声系统的竞技场满满地都是员工以及他们的家人、朋友,打造了一个独特而有活力的氛围。

Information Technology
信息技术

PROJECT LOCATION 项目地点	Moscow, Russia 俄罗斯莫斯科	DESIGNERS 设计师	Arseniy Borisenko, Peter Zaytsev 阿尔谢尼·鲍里先科、彼得·扎伊采夫	DESIGN COMPANY 设计公司	za bor architects 扎·伯尔建筑事务所
PROJECT AREA 项目面积	5800 m²	PHOTOGRAPHERS 摄影师	Maria Turynkina, Dmitry Kulinevich 玛利亚·图林基纳、德米特里·库林叶维奇		

Yandex Stroganov
Yandex斯特罗加诺夫办公室

▶ CORPORATE CULTURE 企业文化

Yandex is one of the largest internet companies in Europe, operating Russia's most popular search engine and its most visited website. Yandex has leading idea to make mostly informal and creative atmosphere that these firms are willing to build up for the staff, because working environment is one of the key factors that affect the company's attraction.

Yandex是欧洲最大的互联网公司之一，运营着俄罗斯最受欢迎的搜索引擎及其访问量最大的网站。由于工作环境也是影响公司吸引力的关键因素之一，所以Yandex的主要想法是创建一种公司乐于为员工搭建的非正式但极具创意的氛围。

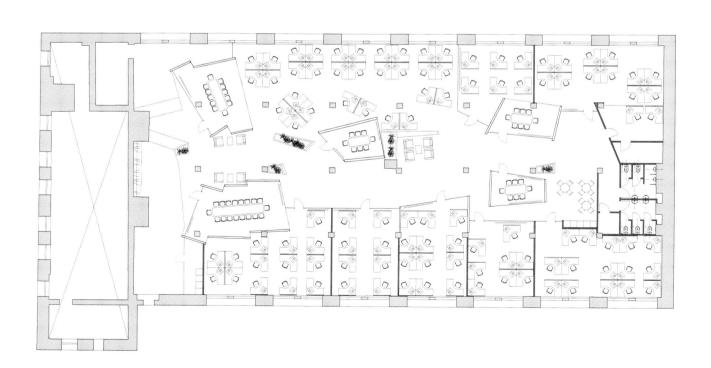

· 111 ·

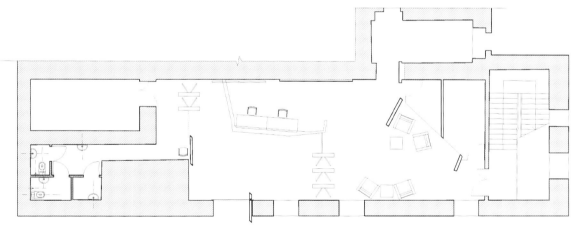

▶ DESIGN CONCEPT 设计理念

This office is located in reconstructed XIX century building. It is full of columns and interstorey premises, which influenced the interiors a lot. The client, wanted to see a "happy" and comfortable interior space.

The first three floors are connected with a generic element, that is intended to form a giant ribbon, that, while penetrating floors, forms streamlined volumes of meeting and conference rooms. The fourth and fifth floors are constructed in a totally different "loft" almost "industrial" style with only two bright elements–large meeting rooms finished with red and yellow color.

这间办公室位于一栋改建过的19世纪大楼，布满圆柱及层间办公场所，深深地影响了空间内部。客户希望看到一个"愉悦"且舒适的室内空间。

前三层楼以一种通用的元素连接，打算形成一条巨大的丝带，贯穿楼层的同时，构成流线型会晤室以及会议厅。四楼和五楼以完全不同的"loft"形式建造，几乎是"工业化"风格，只有两个明亮的元素——装点着大型会议室的红色和黄色。

▶ FEATURE 特色

The first three floors have the following common elements of all Yandex offices, as open communication lines on the ceiling, unique ceiling lights in complex geometrical boxes, and compound flowerpots with flowers dragging on to the ceiling. Alcove sofas by Vitra are used as bright color spots, and places for informal communication.

In the fourth and fifth floors, you may notice two signature elements of za bor architects here–large meeting rooms–architects call them "bathyscaphes", and employees named them "Orange" and "Tomato" due to their colors.

Such difference in decoration is determined with very complex construction elements and level differences in the building (the ceiling height varies from 2 to 6 meters), balconies, beams that were left from the previous tenants. Nevertheless, here we can see new colors, partition walls and flooring. Here, in these neutral grey-white interiors, rather than elsewhere, there are many workplaces completed with Herman Miller systems, and the largest open-spaces. Also there are cafeteria and game room with a sport corner.

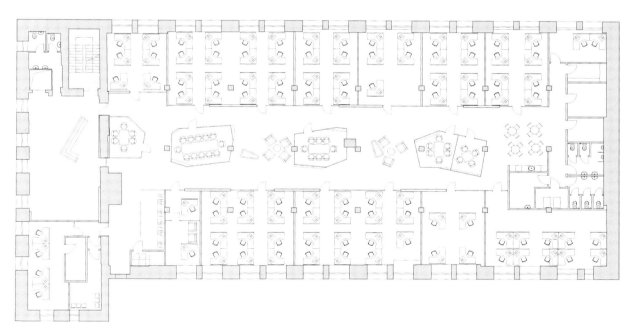

前三个楼层拥有Yandex所有办公室的通用元素，如天花板上的开放式通讯线路、复杂几何状盒子里独特的天花饰灯以及花朵蜿蜒至天花的复合花盆。维特拉的凹室沙发鲜艳明亮，作为非正式交流区之用。

在第四和第五层楼，你能看到扎·伯尔建筑事务所的两个签名元素——大型会议室，设计师称其为"深海潜艇"，员工们则根据颜色把它们叫作"橙子"和"番茄"。

装修上的这些不同取决于该建筑内非常复杂的施工要素、标高差距（层高2~6m不等）、阳台以及之前租户遗留下来的梁柱。然而，我们可以看到新的色彩、隔断墙和地板。在这里，比其他地方更中性灰白的室内空间，有很多配置了赫曼·米勒系统及最大化开放空间的工位，还有自助餐厅和带有运动区的游戏室。

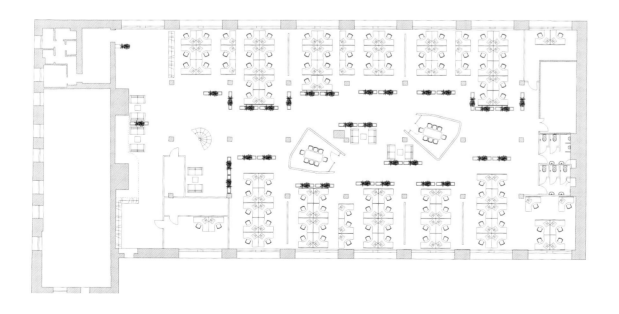

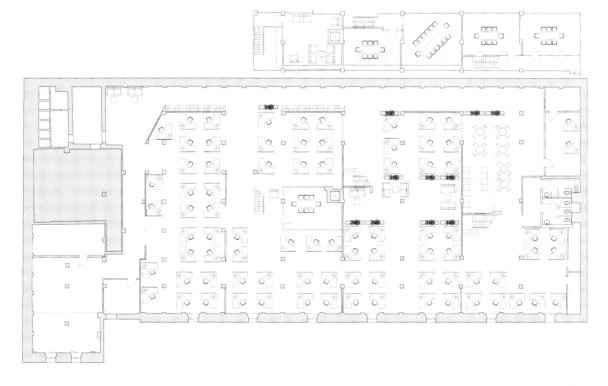

Information Technology 信息技术

PROJECT LOCATION 项目地点	Singapore 新加坡	DESIGN COMPANY 设计公司	SCA Design Pte Ltd (A Member of the ONG&ONG Group) SCA设计公司（ONG&ONG团队成员）
PHOTOGRAPHER 摄影师	Jaume Albert Marti 黎梅·艾伯特·马蒂	PROJECT DIRECTORS 项目总监	Chrisandra Heng, Brandon Liu 王静仪、刘俊雄

Booking.com Office
Booking.com办公室

▶ CORPORATE CULTURE 企业文化

Booking.com B.V., part of the Priceline Group (Nasdaq: PCLN), owns and operates Booking.com™, the world leader in booking accommodations online. International hotel booking site, Booking.com's Singapore office promises to be an inspiration to its employees and clients alike. Given its line of business, the company deals with various destinations and cultures, of which not single one could adequately represent the entire operation as a whole.

Booking.com B.V.公司隶属于Priceline Group集团（纳斯达克上市公司：PCLN），拥有并经营Booking.com品牌，是全世界最大的网上住宿预订公司。作为一家国际酒店预订网站，Booking.com新加坡办事处对其员工及客户承诺这将会成为一个灵感所在。鉴于其业务范畴，该公司涉及众多不同的地区及文化，而其中任何的单一项都不能够充分代表整体运作。

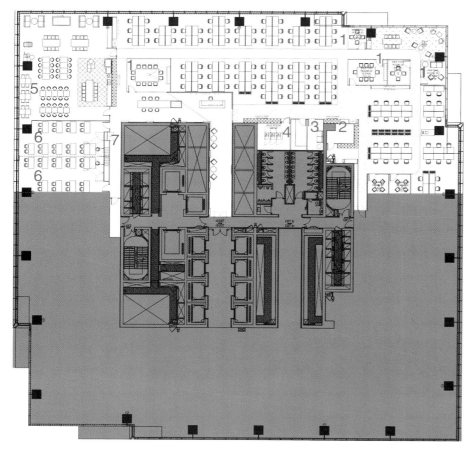

FLOOR PLAN 平面图

1. MEETING ROOM   1. 会议室
2. STORE ROOM   2. 储藏室
3. PRAYER ROOM   3. 祷告室
4. IT ROOM   4. 信息技术室
5. PANTRY   5. 餐具室
6. TRAINING ROOM   6. 培训室
7. PEO SCREEN   7. 聚氧乙烯屏幕

▶ DESIGN CONCEPT 设计理念

The design concept for Booking.com's Singapore office embraces its diversity by theming areas of the office according to different local places and assigning them fun tag lines. The idea was to create a vibrant workspace that could be both interactive as well as socially engaging.

　　Booking.com新加坡办事处的设计理念根据不同的区域依照主题分类，同时赋予它们有趣的宗旨，体现了办公室的多样性。旨在打造一个生机勃勃的工作空间，既有互动性又可以拥有社交魅力。

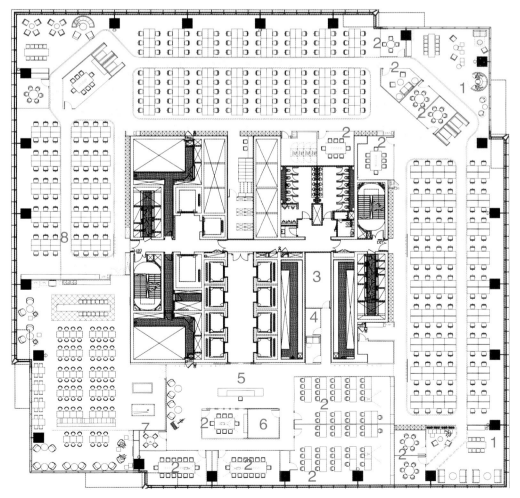

FLOOR PLAN 平面图

1. DESTINATION TOUCHDOWN /COLLABORATION SPACE
2. MEETING ROOM
3. IT ROOM
4. PRAYER ROOM
5. CUSTOMER SERVICE /CREDIT CONTROL RECEPTION/HOLDING AREA
6. STORE ROOM
7. INTERVIEW ROOM
8. CUSTOMER SERVICE

1. 目的地 / 合作空间
2. 会议室
3. 信息技术室
4. 祷告室
5. 客服中心 / 信用管理接待室 / 等候区
6. 储藏室
7. 面试室
8. 客服中心

▶ FEATURE 特色

There are different spaces for different activities, such as learning, working, eating or playing. Despite the demarcation of space, connectivity and interaction were key requirements for the client. In order to meet these specifications, collaboration hubs were placed in highly visible and central areas within the office. Additionally, the office was designed to maximise natural light entering the workplace, while the layout was left flexible enough to allow for the company's future expansion plans.

With its quirkiness as well as its functional design, this is a memorable office where both employees and visitors can feel at ease.

不同的活动对应不同的空间，比如学习、工作、餐饮或娱乐。尽管空间划定了界线，连通和互动仍然是客户的关键需求。为了满足这些具体要求，在办公室的显眼处和中心区域设置了一些协作中心。此外，该办公室的设计试图让自然光线最大限度地进入办公场所，同时布局也能足够灵活，以便实现公司未来的拓展方案。

这是一个令人难忘的办公室，其设计出人意料而又实用，在这里，员工和访客都能够感到轻松安然。

Information Technology 信息技术

PROJECT LOCATION 项目地点	Budapest, Hungary 匈牙利布达佩斯	DESIGNERS 设计师	David Drozsnyik, László Ördögh, Máté Attila Tóth, Oszkár Vági 大卫·多罗兹尼克、拉斯洛·奥多、马特·阿提拉·托特、奥斯扎尔·瓦吉		
PROJECT AREA 项目面积	200 m²	PHOTOGRAPHER 摄影师	Attila Balázs 阿提拉·巴拉兹	DESIGN COMPANY 设计公司	Graphasel Design Studio Ltd Graphasel设计事务所

Google Budapest SPA Office
谷歌布达佩斯温泉办公室

▶ CORPORATE CULTURE 企业文化

Google is an American multinational technology company specializing in Internet-related services and products. These include online advertising technologies, search, cloud computing, and software. Although Googlers share common goals and visions for the company, they hail from all walks of life and speak dozens of languages, reflecting the global audience that they serve. And when not at work, Googlers pursue different interests. Their offices and cafes are designed to encourage interactions between Googlers within and across teams, and to spark conversation about work as well as play.

Google公司是一家美国的跨国科技企业，专门从事互联网相关产品与服务。业务范围涵盖在线广告技术、互联网搜索、云端计算、软件等领域。虽然Google员工拥有共同的公司目标和愿景，但却有着不同的生活背景、说着几十种不同的语言，代表着他们服务的全球受众群体。工作之余，Google员工

有着各种各样的兴趣爱好。他们的办公室和咖啡厅为Google员工进行组内交流和跨组交流提供了便利，大家在休闲娱乐的同时，还可进行工作方面的交谈。

FLOOR PLAN 平面图

1. CUPBOARD	7. BEACH BAR	1. 橱柜	7. 沙滩吧
2. MASSAGE ARMCHAIR	8. BAR	2. 按摩椅	8. 吧台
3. WOODEN SEAT	9. LOUNGE	3. 木质座椅	9. 休息室
4. SAUNA MEETING ROOM	10. POSITION LAMP	4. 桑拿会议室	10. 指示灯
5. STEAM CABIN MEETING ROOM	11. GREEN WALL	5. 汗蒸会议室	11. 绿墙
6. OFFICE	12. MEETING ROOM	6. 办公室	12. 会议室

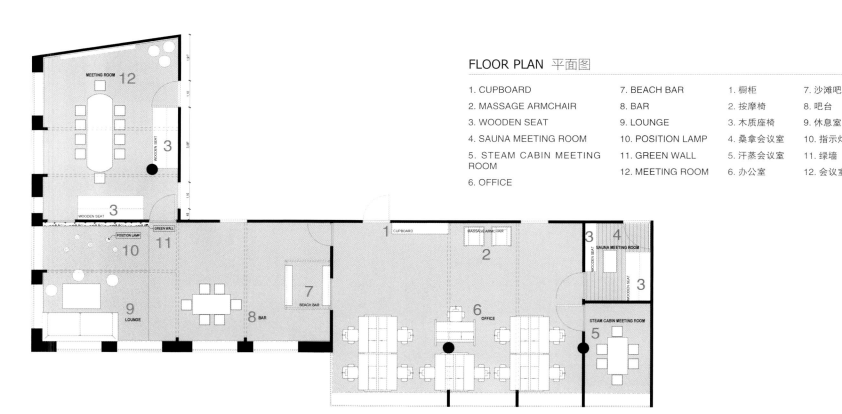

▶ DESIGN CONCEPT 设计理念

The keyword 'SPA' has been chosen for the office in Budapest, since more than 500 springs are present in the Hungarian capital city, which justifies its reputation as a spa town. Nowadays, many traditional and modern baths are available throughout Budapest, of which the two most important ones, the Széchenyi and Gellért baths are represented in the interior design of the office. Whilst during the Roman Empire several spas were located at the centre of the region which was then called Pannonia, and other spas were continuously being added during the Turkish occupation, the lidos–outdoor bathing beaches–had only been created and completed in the first half of the twentieth century.

"温泉"被选为布达佩斯办公室的关键词，匈牙利首都有超过500眼温泉，称其为"温泉之城"名副其实。如今，布达佩斯全城都可以看到很多传统和现代的浴室，最重要的两个分别是切尼温泉和盖勒特温泉，在这间办公室的室内都有显现。罗马帝国时期，当时称作潘诺尼亚地区的中部建立了一些温泉，其他温泉则在土耳其统治时期陆续增加，户外沐浴沙滩仅在20世纪前半叶创建并完成。

Designers could not disregard the importance of Hungarian water polo during the elaboration of the concept, as they as a Nation are pride of the nine Olympic gold medals and three gold medals in water polo world championships, and of course many other great achievements. Accordingly, water polo has become part of the Hungarian bath culture over the years. They aimed to demonstrate parallelism between sports and the spirit of competition in the world of business. As the spas are home to both swimming pools and outdoor rest areas, it was obvious that each room within the office was devoted to a different topic. Because of the extreme diversity of baths, a sauna and a steam bath, as well as a water polo arena and an outdoor beach are also included in the concept.

设计师们不能忽视匈牙利水球运动在概念上精心设计的重要性，因为他们是一个以九枚奥运金牌、三枚水球世锦赛金牌及其他重大奖项而自豪的民族。于是，水球运动成为多年来匈牙利沐浴文化的一部分。他们旨在显现运动与商界竞技精神的相似之处。温泉是游泳池和户外休息区共同之源，显然，办公室内的每个房间都致力于不同的主题。由于沐浴的极大差异性，本概念中还包括了桑拿、蒸汽浴、水球竞技场和户外沙滩。

▶ FEATURE 特色

The requirements set by Google were not only related to the visual appearance, but also insisted on using cheaper recycled materials; so designers incorporated pallets and other residual materials from constructions. They also tried to "smuggle" the works of contemporary Hungarian designers into the small details, combined with antique furniture and tables characteristic of old baths.

谷歌的要求是，不仅要与视觉外观有关，而且要坚持使用低成本的回收材料；所以设计师们融合了踏板以及其他建筑剩余材料。同时，他们还"偷运"当今匈牙利设计师们的作品，将它们运用在一些小细节上，并结合了古董家具与带着古老浴室特色的桌子。

The original recessed ceiling fluorescent lights used to overlight the premises. By eliminating the suspend ceiling the conduits and cables became visible, which was then supplemented by further conduits painted to the color of the Google logo. Thus the perception of height increased, and the many colorful ceiling pipes created a slightly industrial atmosphere. Whilst the more interesting regions are emphasized using spotlights, designers aimed to achieve an overall cozy gloom in the area.

原有的嵌入式天花荧光灯过去常令办公场所光度过强。通过消除悬浮式天花的感觉，让导管和电缆裸露，辅以进一步的管道喷涂，涂成谷歌商标的颜色。这样，让人感觉层高得到了增加，各种彩色的天花管线打造了轻度工业化的氛围。同时，更有趣的区域通过聚光灯进行强化，设计师的目标是在这里实现一个整体舒适的幽暗之感。

Printed wallpaper surfaces are strongly emphasized in the new office. At some places, however, the printed reproduction of textures would not have been sufficiently authentic, so the sauna boardroom is made of real wood, and some parts of the mosaic are made of original materials. Designers insisted on using a lot of real green plants because the microclimate of living plants exerts a positive impact on the well-being of the employees.

新办公室内极力凸显了印花壁纸。然而在一些地方，一些材质的印花复制品并不是特别正宗，所以桑拿房由实木及原始材料质地的马赛克制成。因为活体植物能够对员工幸福感有积极的影响，所以设计师们坚持使用大量真实的绿色植物。

It was important to allow for continuous complementation and improvement by the employees of the office. Therefore, the appearance of colorful towels or beach balls in the office is just as mundane as logging on to a video conference in a swimming hat.

允许办公室员工持续补足和提升自己是很重要的。因此，在办公室中出现彩色毛巾、水球、员工们戴着泳帽登录视频会议的场景都是很平常的。

Information Technology
信息技术

PROJECT LOCATION 项目地点	Santa Monica, California, USA 美国加利福尼亚州圣塔莫尼卡	DESIGNERS 设计师	Deniece Duscheone (Principal Designer), Christopher Maresca (Principal Architect) 丹尼斯·杜舍恩（主设计师），克里斯多夫·马雷斯卡（主建筑师）		
PROJECT AREA 项目面积	9250 m²	PHOTOGRAPHER 摄影师	Christopher Stark Photography 克里斯托弗·斯塔克摄影工作室	DESIGN COMPANY 设计公司	SKIN Design Studio SKIN设计事务所

Cornerstone OnDemand
Cornerstone OnDemand办公室

▶ CORPORATE CULTURE 企业文化

Cornerstone OnDemand is an international technology company that specializes in the development of talent management software. It is also currently a leader in the Los Angeles, CA technology development sector. They are focused on innovative and creative environments that will allow a multi-generational cross pollinating of ideas in the work place. They have a strong belief in not only the community of their own work place, but in the larger context of creating integrated environments on local and global platforms.

Cornerstone OnDemand公司是一家专门研发人才管理软件的国际化科技公司。它也是当前洛杉矶技术发展部门的领导者。他们注重创新型的环境，这样在工作区域内便会有来自不同年代的思想汇聚。不仅在他们每个人工作的地方，而且在创造当地及全球平台集成化环境的大背景下，他们都拥有强烈的信念。

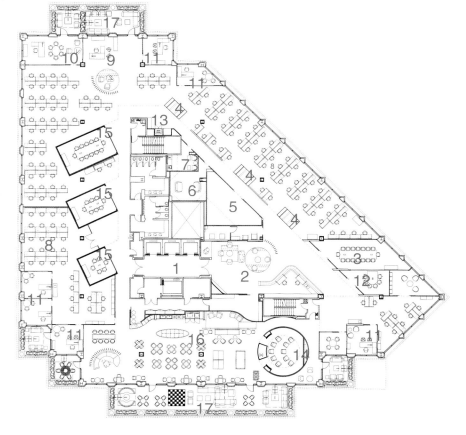

FLOOR PLAN 平面图

1. ELEVATOR LOBBY	1. 电梯门厅
2. RECEPTION	2. 接待室
3. BOARD ROOM	3. 报告厅
4. COLLABORATIVE SEATING	4. 联排座椅
5. YOGA ROOM	5. 瑜伽室
6. MEDITATION ROOM	6. 冥想室
7. MATERNITY ROOM	7. 妇科诊室
8. OPEN WORK DESKS	8. 开放区办公桌
9. CEO RECEPTION	9. 首席执行官接待室
10. CEO OFFICE	10. 首席执行官办公室
11. OFFICE	11. 办公室
12. INTERNET LOUNGE	12. 网络休息室
13. SNACK AREA	13. 点心区
14. MEDIA THEATER	14. 媒体区
15. CONFERENCE ROOM	15. 会议室
16. CAFE/MULTI-PURPOSE ROOM	16. 咖啡室/多功能室
17. OUTDOOR LOUNGE	17. 户外休息室

▶ DESIGN CONCEPT 设计理念

The existing client space was a large compartmentalized, isolating, grouping of spaces lacking in any natural light and which did not meet Cornerstone OnDemand's vision for the company's philosophy and culture of reimaging work, enhancing productivity, and connectedness.

SKIN Design Studio solved the clients objectives of creating a space that reimagines how people work, enhances productivity, and connectedness by exploring how their experience in luxury hospitality design could be used to solve these objectives. By bringing the design ideas of how a hotel operates on a guest services platform they were able to investigate the integration of hospitality philosophies into the corporate environment. This lead SKIN to designing the client space as an extremely inviting environment with large light filled open spaces. Further cues that SKIN cross bred from hospitality environments were the creation of multiple lounge type spaces such as the reception area, meditation room, and café media theater. These unique spaces helped to create a diverse work environment for employee collaborations with the attempt to unify connections and empower people.

To manage the challenge of a rapidly expanding client company during design and construction it was important that the entire project team remained flexible and focused on great communication and interaction with the client. This allowed the project team to stay on top of the continued design changes as required due to the client's growth and the additions of many employees.

原有的空间是一个大型、多区域、孤立的群组空间，缺乏自然光线，并不符合Cornerstone OnDemand公司在重塑形象、提高生产率及联结性方面的经营理念及文化设想。

SKIN设计事务所达成了客户的目标，通过探索如何运用他们在豪华酒店的设计经验来解决这些问题，打造了一个可以重新想象人们工作模式、提高生产率及联结性的空间。通过引入酒店待客服务平台运作模式的设计灵感，他们便可以探究待客哲学与企业氛围的融合。为此，SKIN设计事务所将该空间打造成一个极具魅力、光线充足的大型开放式空间环境。SKIN设计事务所融入酒店式的环境还可以进一步在各种休闲空间的设计上体现出来，例如接待室、冥想室以及咖啡媒体室。这些独特的空间为员工协作创造了一个多样化的办公环境，旨在加强团结，授予员工更多权利。

为应对该公司迅速发展所面临的挑战，在设计建造过程中，整个项目团队保持灵活性，重点关注与客户的交流和互动尤为重要。如此，设计团队也能在客户公司不断发展及增加员工时，随时根据需求变更设计。

▶ FEATURE 特色

The client project was constructed in an existing building that currently has a LEED O+M: Existing Buildings Gold Certification. The building landlord further requires that LEED-EB certification documents be completed for each tenant improvement project within the building. This led the design and construction team to follow smart practices regarding the specification and application of materials, such as low-VOC paints compliant with the Green Seal GS-11 standard. For the sustainable design objectives it was important for SKIN to open up the existing space as much as possible. This led to using a minimum of dividing partitions and ample amounts of glass walls to allow natural daylighting to penetrate deep into the building core and reducing the need for artificial lighting.

该项目建造于一栋经过绿色建筑运营和维护评估、获得既有建筑黄金认证的大楼内。大楼的主人要求每个租户在整修过程中都应完全符合LEED-EB文件的要求。这使得设计和施工团队在规格和材料应用方面遵循明智的方案，例如使用了符合美国绿色认证（Green Seal GS-11）的低挥发性涂料。为达成可持续设计目标，尽可能地开发现有空间对SKIN设计公司来说非常重要。因此他们运用最少的隔断、充裕的玻璃幕墙来营造自然光线，使其深深地穿过建筑核心，减少人工照明的需求。

Information Technology
信息技术

PROJECT LOCATION 项目地点	Yokohama City, Japan 日本横滨市	DESIGNERS 设计师	Masahiro Yoshida, Riyo Tsuhata 吉田昌弘、津秦理世	DESIGN COMPANY 设计公司	Kamitopen Architecture-Design Office Co., ltd. Kamitopen建筑事务所
PROJECT AREA 项目面积	143 m²	PHOTOGRAPHER 摄影师	Keisuke Miyamoto 宫本启介	ARCHITECTURAL COMPANY 建筑公司	Tub Studio Tub工作室

Yudo Office
柳道办公室

▶ CORPORATE CULTURE 企业文化

"Connection"
Yudo Ltd is a group of creators that gives people enjoyment and happiness by providing services, which excites their five senses, mainly on smartphones.
Just by using your index finger, you can play music or draw a picture.
Talk with someone who lives far away.
All of the staffs excitingly develops and sends out their services to the earth, which is in the solar system of the huge outer space, by engaging their brains.

"连接"
柳道有限责任公司是一群创造者，主要通过提供智能手机方面的服务刺激人们的五感，带来享受和幸福。仅仅动动食指，你就可以播放音乐或画一幅画。
还可以与不在身边的人聊天。
全体员工运用头脑振奋人心地研发服务项目，并向整个地球输出。

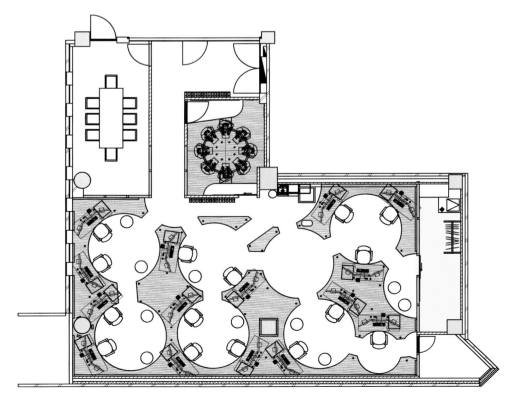

▶ DESIGN CONCEPT 设计理念

This time, the request of Yudo Ltd was to make an office in which the staffs can engage their brains more easily. In other words, they wanted the designer to make a space where they can feel the "connection" of the other staffs with their five senses.

  这一次，柳道有限责任公司的要求是设计一间能够让员工更方便地开发大脑的办公室。换言之，他们希望设计师打造一个空间，让人们通过五感，感受到彼此之间的"连接"。

▶ FEATURE 特色

Masahiro, therefore, suggested to cover the whole room with one board and create holes that people can get in. As a result, desks of all the staffs got connected as one like a lane in a factory, "connecting" the works and minds of the staffs.

Moreover, Masahiro had designed the entrance so that the guests and staffs get "connected" by an analog system using a speaking tube. This is also an approach to hand on the warmth of a human being to Yudo office, a company that offers digital services.

Masahiro hopes that this office will allow the staffs of Yudo to connect stronger with themselves as well as with all the human beings in the world through the services that are created here.

因此，吉田建议用一块板子覆盖整个空间，创建一些人们可以进入的孔洞。结果，所有员工的办公桌都被串联起来，像一间工厂的产业链一般，"连接"员工之间的工作与心灵。

此外，昌弘还设计了入口，用一条传话管模拟系统使访客和员工间建立"连接"。在提供数字化服务的柳道办公室内，这也是传递人性化温暖的一种方式。

通过在这里创造的服务，吉田希望，这间办公室可以让柳道的员工与他们内部及整个外部世界的人类有更加紧密的联系。

Information Technology 信息技术

DESIGNER 设计师	Shirli Zamir 雪莉·扎米尔	DESIGN COMPANY 设计公司	Setter Architects 赛特建筑
PHOTOGRAPHER 摄影师	Uzi Porat 乌兹·波拉特	PROJECT MANAGER 项目经理	Chen Yaron (Yaron-Levy) 陈亚伦（亚伦-莱维）

Autodesk Israel – Tel Aviv
欧特克以色列特拉维夫办公室

▶ **CORPORATE CULTURE 企业文化**

Setter Architects has recently completed a new cutting edge development center in Tel Aviv for Autodesk, an international leader in 2D and 3D design software. The new office extends over four floors of a new tower on Tel Aviv's prestigious Rothschild Boulevard. Autodesk defines itself not just as a software company but also as a "design" company and ascribes tremendous importance to the appearance and functionality of its offices.

欧特克公司是世界领先的二维、三维设计软件公司，赛特建筑最近在特拉维夫市为其完成了一间新的前沿发展中心。新办公室占据一栋新式高楼的四个楼层，该高楼位于特拉维夫市享誉盛名的罗斯柴尔德大道。欧特克公司不仅将自身定义为一家软件公司，而且以"设计"公司自居，它还非常注重办公室的形象和功能性。

▶ DESIGN CONCEPT 设计理念

Setter Architects was tasked with designing a space to reflect a creative blend of Autodesk's corporate and local cultures, which puts a strong emphasis on the well-being of its employees—even to the point of allowing staff to bring their dogs to work! With a planning directive to bring pleasure and fun to a flexible work environment, Setter was also tasked with creating a design that would highlight Autodesk's culture of collaborative teamwork, open communication between teams, and creating spaces that would work for both alone and in groups.

The project team embarked on an engaging process of capturing the needs and expectations of both local managers and executives from the company headquarters in San Francisco. This dialog resulted in the use of rich three-dimensional design and a broad range of materials, applying the language of software programs that Autodesk develops in its design and graphics, and creating a truly pleasurable work setting.

赛特建筑被委任设计一个可以展现欧特克公司企业文化与当地文化创意交融的空间，着重强调员工的幸福感——甚至达到可以允许职员携带宠物狗来工作的程度！本着将欢乐与趣味性带入灵活工作环境的设计目标，赛特建筑还要负责打造一个可以突出欧特克公司企业文化的设计，即团队协作与团队间的开放式交流，并创建可供独立抑或分组工作的空间。

项目团队首先着手于了解当地管理者与旧金山总部高管们的需求和期望。交流的结果是使用丰富立体化设计及广泛材料，并且应用欧特克公司在其设计与图像研发中的软件程序语言，打造一处真正富于乐趣的办公环境。

▶ FEATURE 特色

A special feature of the new development center is to take advantage of the magnificent 360° surrounding city view of the beautiful modern Tel Aviv. With that in mind, the space was designed for all employees to enjoy the views from every corner of the office space. The interior design also includes an assortment of simple and raw industrial-style materials, from concrete floors to rusted treated metal walls, reclaimed wood from discarded window frames and plaster, and vibrant wallpaper with graphic elements infused with Autodesk branding. The intriguing blend of these elements creates a unique sense of flow.

The office plan offers ample quiet work-areas, informal collaboration areas, and formal conference rooms, which also serve as a visual and acoustic barriers in the transition from public spaces and work areas. These rooms are easily accessible from the work areas and the public spaces, to maximize employees' options on how they would like to work and meet.

Sustainability at Autodesk is a company-wide initiative involving all aspects of Autodesk's business. The project was planned and designed in accordance to the guidelines of U.S. Green Building Council. The office has a sophisticated monitoring system controlling the lights, drapes, and air-conditioning, which continually monitors and tests the air quality in most enclosed rooms. The project is registered under LEED® Green Building Certification program pursuing the highest rating of LEED® Platinum in Commercial Interiors.

这间新发展中心的一大特色便是利用了美丽现代的特拉维夫市周边360度壮丽城景。有鉴于此，该空间的设计是为了让所有员工在办公室的每个角落都能观赏到美景。室内的设计还包含了各式各样简单、原生态的工业化材料，从混凝土地板到防锈金属墙壁，从再生木制品到废弃窗框和灰泥，以及带有欧特克公司品牌图形元素的充满生机的壁纸。这些元素的有趣融合打造了一种与众不同的流动感。

办公室的规划提供了充裕的安静办公区、非正式洽谈区以及正式会议厅，它们同时充当了公共区及办公区过渡中的视觉、听觉屏障。员工可以轻易地从办公区或公共区进入这些空间，这样他们就可以最大化地选择如何工作及会面。

可持续性是欧特克全公司的首创精神，涵盖其业务的各个方面。该项目根据美国绿色建筑委员会的指导方针进行规划设计。该办公室拥有复杂的监控系统，可以控制灯光、窗帘及空调，可以在大多数封闭的房间内持续监控并检测空气质量。该项目已在绿色建筑认证计划下登记，旨在获得商业空间设计中最高级别的绿色建筑铂金奖。

Information Technology
信息技术

PROJECT LOCATION 项目地点	Kiev, Ukraine 乌克兰基辅	DESIGN COMPANY 设计公司	Soesthetic Group Soesthetic团队
PROJECT AREA 项目面积	3000 m²	PHOTOGRAPHER 摄影师	Alex Pedko 亚历克斯·派德科

Office Playtech
Playtech办公室

▶ CORPORATE CULTURE 企业文化

Playtech is the world's largest online gaming software supplier, offering cutting-edge, value added solutions to the industry's leading operators. Playtech's Kiev team is continuously growing and increasing with young talented specialists.

Offices in each country have their own design with its own corporate culture but the best habits from offices around the world Soesthetic Group adopted for this new office. For example there is large meeting area in the central part of the office where you could collect an audience for a meeting in couple of minutes. In each area whether it's work or leisure Soesthetic Group designed graphics symbols and navigation. It helps to follow the basic social rules and develops better corporate culture.

Playtech是全球最大的在线游戏软件供应商，为该产业的领头羊提供前沿、有附加价值的解决方案。Playtech基辅团队随着年轻、有才华的专家加盟而逐步成长扩张。

这些国家的办公室都有基于其企业文化的特色设计，然而Soesthetic团队

在打造这间办公室的时候融入了自身的良好习惯。比如，该办公室中心位置有一间大型会议室，你可以在几分钟之内找到人来开会。不论是工作区还是休闲区，Soesthetic团队都设计了图形符号及导航。这有助于遵循基本社交规则，发展更好的企业文化。

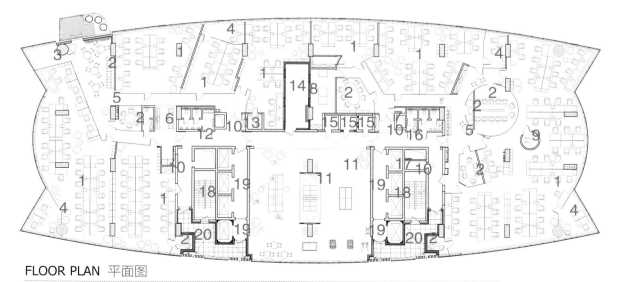

FLOOR PLAN 平面图

1. OFFICE SPACE
2. MEETING ROOM
3. RELAX ROOM
4. OFFICE
5. CORRIDOR
6. UNISEX WC
7. STORAGE
8. LAN SWITCH ROOM
9. AUDIO BOOTH
10. FIRE SAFE CORRIDOR
11. RECREATION ZONE
12. MEN'S WC
13. HVAC ROOM
14. WENT-CHAMBER
15. ELECTRIC
16. WOMEN'S WC
17. TECHNICAL
18. STAIRCASE
19. ELEVATOR HALLWAY
20. FIRE SAFE LOGGIA

1. 办公空间
2. 会议室
3. 休息室
4. 办公室
5. 走廊
6. 男女通用洗手间
7. 储藏室
8. 局域网交换机室
9. 音频亭
10. 消防通道
11. 娱乐区
12. 男士洗手间
13. 空调系统室
14. 停留室
15. 电机室
16. 女士卫生间
17. 技术室
18. 楼梯间
19. 电梯门厅
20. 消防门廊

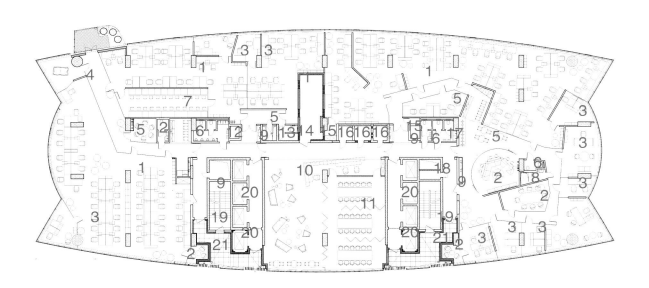

FLOOR PLAN 平面图

1. OFFICE SPACE	8. MANAGER'S WC	15. STORAGE	1. 办公空间	8. 管理人员洗手间	15. 储藏室
2. MEETING ROOM	9. FIRE SAFE CORRIDOR	16. ELECTRIC	2. 会议室	9. 消防通道	16. 电机室
3. OFFICE	10. RELAX ZONE	17. WOMEN'S WC	3. 办公室	10. 休息区	17. 女士洗手间
4. RELAX ROOM	11. TRANSFORMER CONFERENCE ROOM	18. TECHNICAL	4. 休息室	11. 变压会议室	18. 技术室
5. CORRIDOR	12. MEN'S WC	19. STAIRCASE	5. 走廊	12. 男士洗手间	19. 楼梯
6. UNISEX WC	13. HVAC ROOM	20. ELEVATOR HALLWAY	6. 男女通用洗手间	13. 空调系统室	20. 电梯门厅
7. GAME TERMINAL ROOM	14. SHOWER ROOM	21. FIRE SAFE LOGGIA	7. 游戏终端室	14. 淋浴室	21. 消防门廊

▶ DESIGN CONCEPT 设计理念

Playtech's Kiev office is located in the new contemporary business centre with a panoramic view along the entire perimeter of the building. The starting point was to design complex floor plan with numerous work areas, relax and social zones. It needed to maximise the use of daylight and open the stunning 22nd floors panoramic view. This was a challenge due to the extremely poor planning of the building central core. All the elevator shafts, stairs and engineering channels were designed according to residential guidelines. To make matters worse, they were all cast in concrete, so there was nothing to be done with them. This basically split the floor plan into three separate parts with only a cramped hallway connecting them.

Yet despite all that Soesthetic Group decided to preserve the architectural image and make transparency and openness its main feature. Almost all existing brick wells were demolished to bring light into the hallway. They placed the reception, relaxation and activity into the bright central area, and all the spread working zones around the perimeter, to give each workplace a panoramic view. Coffee points and other public amenities were placed in between the two hallways to bring life into the central core. Existing concrete structures and white colored walls became the background material. Glass structures with semi-transparent pattern and color accents completed the overall perception. The view became visible from most parts of the office, the plan fits in a lot of work stations and leisure areas without overcrowding the internal air.

Playtech基辅办公室坐落在一栋新的现代化商业中心，将该栋大楼的全景尽收眼底。第一步是设计数不胜数的办公区、休闲区和社交区的复杂平面图。需要将日光的利用最大化，打开22楼那极好的全景图。大楼核心位置的规划不佳，这是一项很大的挑战。所有升降机井、楼梯、工程类通道都是按照住宅指导方针设计的。更糟糕的是，它们都不可移动，所以什么也做不了。这使得建筑的平面图基本被划分为三个不同的部分，只有一条狭窄的过道连接着彼此。

尽管如此，Soesthetic 团队决定保持建筑影像，将透明和开阔作为其主要特色。为将光线带入走廊，几乎所有现存的砖井都被拆除。他们将接待处、休息室、活动区域通通带入明亮的中心区域，所有展开的工作区域都围绕着边界，这样每个工作空间都能看到全景。咖啡点和公共设施被安置在两个走廊之间，将生命色彩带入核心区域。

现有的混凝土结构和白色墙壁成为了背景材料。有着半透明图案和着色重点的玻璃结构完成了整体感知。办公室多数地方都可以观景，该计划适合很多工作台以及休闲区域，而不会聚集在内部环境中人满为患。

▶ FEATURE 特色

In this project Soesthetic Group paid particular attention to the public areas: reception, coffee points, relaxation and activity. The office has numerous public zones for people to communicate and to encourage the team's social life. There is a giant game box in the middle of the floor where employees could play and test the company's new products, communicate with each other and relax.

The main element of the activity area become the GAME BOX. It's made of perforated MDF with inner surfaces of black tempered glass. Screens for games are installed behind the glass. All features connect at an accurate 45 degrees to make the complex technological interactions work.

The reception is made of rusted metal with a perforated gradient pattern. Lighting, hidden behind the perforation projects it's pattern onto the floor, the shape of which is similar to cutting on the GAME BOX facades and includes the Playtech logo detail. Soesthetic Group paid attention to graphic navigation so it lets the company employees understand the floor layouts and ease the use of public areas and subjects.

To maximise the room height designers left the concrete ceiling slabs open with visible engineering structures. The double glass walls between the offices makes a comfortable acoustic atmosphere. The use of LED lights helped to keep healthy energy consumption levels.

These solutions made the interior incredibly bright, and transparent and yet cosy at the same time.

Soesthetic团队在这个项目中特别注重公共区域：接待处、咖啡点、休息区以及活动区。该办公室内有许许多多的公共区域，人们可以相互交流、激励团队的社交生活。这层楼的中央有一个巨大的游戏盒子，员工们可以玩乐、测试公司新产品或者跟大家交谈、休息。

活动区的主要元素便是那个游戏盒子。它由穿孔中密度纤维板及黑色钢化玻璃内底板构成。游戏屏幕安装在玻璃后面。所有特性以45度角精确连接，使这个复杂的技术交互运作。

接待室由生锈的金属和穿孔斜梯图案组成。照明设备隐藏在穿孔中，向地面投射图案，形状与游戏盒子外立面非常相似，都含有Playtech的商标细节。Soesthetic团队重视图像导航，让公司员工理解平面布局，缓解公共区域和主题的使用。

为使房间高度最大化，设计师保留了混凝土天花板，让其显示出工程结构的开放性与可见性。各办公室间的双侧玻璃墙打造了一个舒适的听觉氛围。LED灯的使用帮助维持健康能源消耗水平。

这些方案令室内空间极为明亮、通透，同时温馨惬意。

Information Technology
信息技术

PROJECT LOCATION 项目地点	Budapest, Hungary 匈牙利布达佩斯	DESIGNERS 设计师	Krisztina Madi, Aron Lancos, Edit Moder, Bence Torok 克里斯蒂娜·马迪、阿伦·兰科斯、埃迪特·莫德、本斯·托罗克	DESIGN COMPANY 设计公司	MadiLancos Studio Ltd. MadiLancos工作室
PROJECT AREA 项目面积	1200 m²	PHOTOGRAPHER 摄影师	Zsolt Batar 若尔特·巴特		

Ustream
Ustream办公室

▶ CORPORATE CULTURE 企业文化

Ustream is a succesful Hungarian IT startup company. They are working in a very competitive and very fast changing field so having the right mindset and values are very important to remain a successful company. Their core values are: honesty, courage, mastery, respect, ownership and impact. Their culture is team-focused and success-driven.

Ustream是匈牙利一家成功的创业公司。他们从事于一个竞争非常激烈且变化迅速的领域，因此正确的心态和价值观对于公司继续获得成功是非常重要的。他们的核心价值观是诚信、勇气、通达、尊重、所有权和影响力。他们的文化是团队合作和不断进取。

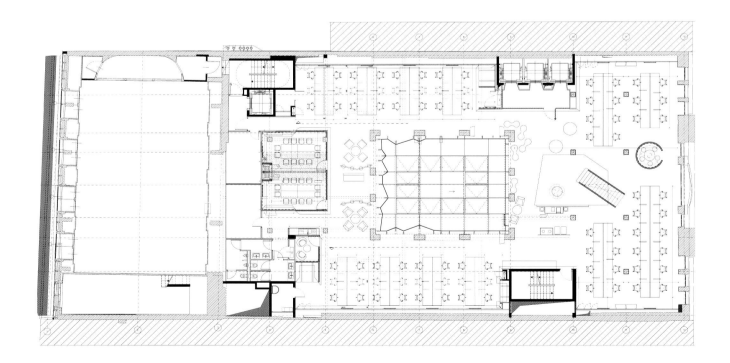

▶ DESIGN CONCEPT 设计理念

Ustream was looking for a large open space office to enhance collaboration and efficiency of work, with adequate place for organising meetups for customers and clients. Their choice fell on a historical monument—a department store built at the end of the 19th century in the heart of Budapest, which was recently refurbished as an office building. Designers' main challange was to keep the industrial look and feel of the building while delivering an office that ensures flexibility and serves the required functions.

Ustream试图寻找一个大型开放的空间作为办公室，以加强合作、提高工作效率，这样就可以为顾客和客户的会面提供适当场所。他们最终选择了一个历史遗迹——19世纪末期在布达佩斯市中心建成的一间百货公司。设计师的主要挑战是在打造一间灵活性与功能性并存的办公室的同时，保持这栋大楼的工业化外观及感觉。

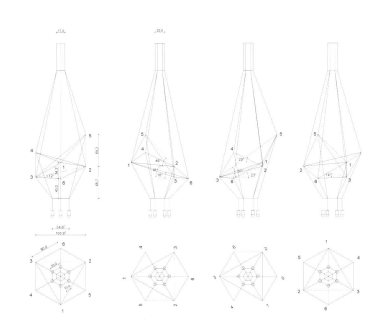

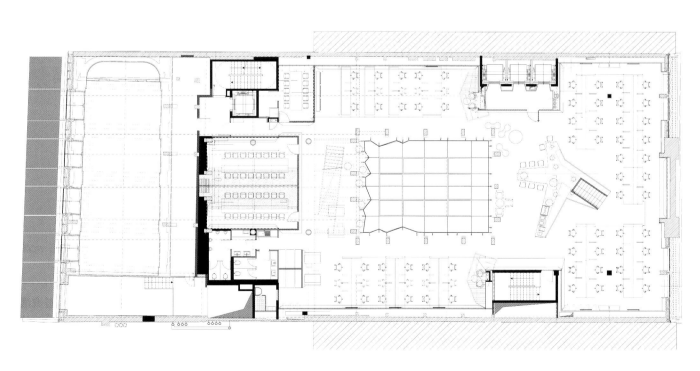

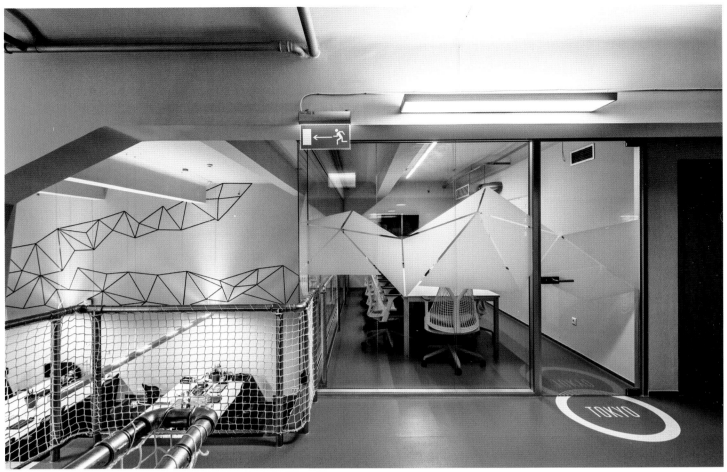

▶ FEATURE 特色

Designers connected the three floors with new internal staircases to facilitate circulation. The layout of the offices is entirely open except for the closed meeting rooms tucked in the darker section of the floorplate. They made a variety of informal meeting and focus points spread out on the floorplates: a bar counter with plants integrated, several booths with accoustic cover, gezebos looking down at the office space.

The top floor which has a spectacular glass roof and a terrace is used as a break-out and dining area. This floor also houses the weekly meetups. The design features industrial elements such as the concrete like vinyl floor and the custom designed chandeliers and graphics which have the same fragmented pattern. The industrial look is softened by the bold colors used on the walls and upholsteries.

设计师将三种色彩与崭新的内部楼梯联结在一起，促进流通性。除了一些蜷缩在楼面较暗角落的封闭会议室外，办公室的整体布局为开放式。他们打造的各种各样的非正式会议及聚焦点遍布楼面各处，例如有着植物的服务台、隔音的房间和办公区的瞭望台。

顶层有壮观的玻璃穹顶和平台，被用作休息及餐饮区域。这一层还有供每周聚会使用的空间。该设计以工业元素为特色，比如混凝土状的塑胶地板、定制的水晶吊灯以及带有相同碎片图案的图形。工业化的外观被墙壁及装饰物上浓重的色彩弱化。

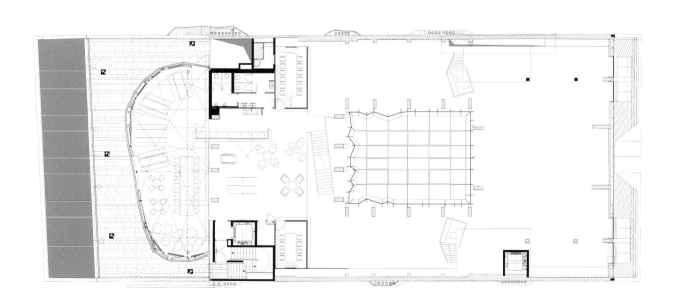

Information Technology 信息技术

PROJECT LOCATION 项目地点	Stockholm, Sweden 瑞典斯德哥尔摩	DESIGNER 设计师	Peter Sahlin 彼得·萨林	ARCHITECTS 建筑师	Mette Larsson Wedborn (Part 1), Mari Owrenn (Part 2) 梅特·拉松·韦德波（第一阶段）、玛丽·欧文斯（第二阶段）
PHOTOGRAPHER 摄影师	Jason Strong Photography 詹森·斯特朗摄影	LIGHTING DESIGN 灯光设计师	Beata Denton 贝亚特·丹顿	DESIGN COMPANY 设计公司	pSArkitektur pS建筑设计公司
ASSISTING ARCHITECTS 助理建筑师	Erika Janunger, Thérèse Svalling (Part 1), Martina Eliasson, Emilie Westergaard Folkersen, Thérèse Svalling (Part 2) 埃里卡·詹纽格、泰雷兹·斯瓦林（第一阶段）、玛蒂娜·埃利亚松、埃米莉·韦斯特加德、福凯尔森、泰雷兹·斯瓦林（第二阶段）				

Skype Office in Stockholm
Skype斯德哥尔摩办公室

▶ CORPORATE CULTURE 企业文化

The Stockholm based architects pSArkitektur have designed the new Swedish office for Skype. It is located in the München brewery in Stockholm-a 19th century red brick industrial complex now transformed into a cluster of offices, conference facilities and venue. pSArkitektur helped Skype find this central and charming location, which attracts talented employees who wouldn't want to work in a bleak suburb. The Skype office houses about 100 employees, mainly engaging in developing audio-and video technology for the Skype platforms. The master plan and the interior design by pSArkitektur reflect the Skype spirit—a joyful, user friendly tool connecting people all over the world. So is the space—playful and functional.

pS建筑设计公司总部设在斯德哥尔摩，刚刚完成了Skype公司在瑞典的新办公室。该址位于斯德哥尔摩的慕尼黑啤酒厂内，一栋19世纪红砖工业中心，现在被改造成一组办公室、会议设施和会场。pSArkitektur帮助Skype找到了这个讨人喜欢的中心位置，吸引了众多不愿在荒凉郊外工作的精英员工。Skype办公室有100名员工，主要为Skype平台研发音频及视频技术。

pSArkitektur的总体规划和室内设计反应了Skype公司的精神——愉快的、亲近使用者的工具连接了世界各地的人们。该空间也是如此，幽默且功能性十足。

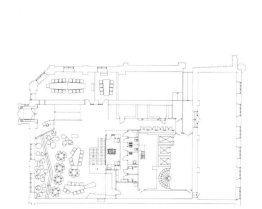
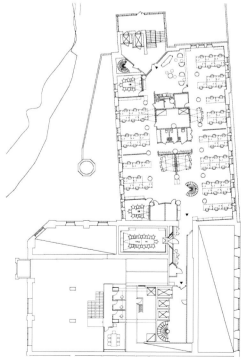

· 173 ·

▶ DESIGN CONCEPT 设计理念

The triangular and rhomboid shapes of the carpet pattern, as well as the design of workplaces, tables and bar, are based on the concept of a net with connecting nodes. The colorful and round shapes of the furniture are derived from Skype's graphic profile. Also, the Skype logo, being a cloud, has been interpreted into cloud-like light fittings around the office—made up of clusters of large white spherical lamps. The office's focus on development of audio and video is manifested in a large number of custom designed wallpapers, in which pictures of headsets and cables, create patterns and bring an individual look to each room.

三角形和菱形的地毯图案，以及工作场所、办公桌、吧台的设计都基于饱含连接节点的网巢概念。彩色圆形家具从Skype图像轮廓中衍生出来。另外，云形的Skype商标被诠释为这间办公室四周的云状灯饰配件，由成串的巨大白色球形灯构成。大量定制壁纸展示出该办公室影音方面的开发重点，耳机和电缆图案创造了众多模式，为每个房间带来独特的面貌。

▶ FEATURE 特色

The design of the Skype office is not only about what is visible to the naked eye. Skype values a playful work environment in which the employees are encouraged to think, play and generate creativity. Instead of the old fashioned offices with cubicles, this office has a mix of regular desks in open office spaces, hot desks, seating areas, meeting rooms, conversation booths, play areas, cafés and silent areas. By organizing these in the right way, social meeting points have naturally appeared in the architecture. This creates a playful atmosphere that allows good, crazy and brilliant ideas to develop.

Finally, the office has been equipped with the newest in technology and it has the best acoustics possible, something that is necessary for this type of work. This is obtained by the use of textiles and a large number of wall mounted sound absorbers with and without prints.

  Skype办公室的设计不仅仅关于肉眼可见的事物。Skype重视轻松幽默的工作氛围，可以鼓励员工思考、娱乐、产生创造力。取代守旧的格子间办公室，这里采用了一种混搭模式，将开放空间内的正规办公桌、公用办公桌、落座区、会议室、对话小隔间、娱乐区、咖啡厅和静音区结合在一起。运用正确的方法组织这些，社交聚会在这栋建筑中自然而然地出现。

  最终，该办公室具备科技领域最新的配置，拥有最佳音响效果，对于这种工作性质来说是非常必要的。这一切得益于纺织品的使用以及有大量印花和没有印花的壁挂式消声器。

Information Technology 信息技术

PROJECT LOCATION 项目地点	Delft, The Netherlands 荷兰代尔夫特	DESIGNER 设计师	Jamie Van Lede 杰米·万·莱德	DESIGN COMPANY 设计公司	Origins Architects Origins建筑事务所
PROJECT AREA 项目面积	400 m²	PHOTOGRAPHER 摄影师	www.stijnstijl.nl		

Fabrique Delft
代尔夫特Fabrique办公室

▶ CORPORATE CULTURE 企业文化

Fabrique specializes in high-end brand identity and digital media. The company is one of the first to adapt to the SCRUM work methods, which requires special attention in the design of the interior and flexible workspaces.

Fabrique公司专营高端品牌标识及数字媒体，是第一批适应SCRUM工作模式的公司之一，这就要求特别关注其室内及灵活办公场所的设计。

FLOOR PLAN 平面图

1. ADMINISTRATION OFFICE
2. MEETING ROOM
3. STAIRCASE
4. ENTRANCE
5. TOILET SPACE
6. INVALID TOILET
7. SHOWER
8. TOILET
9. TECHNICAL INSTALLATIONS
10. SAMPLE ROOM/QUIET SPACE
11. REPRO ROOM/KITCHEN/STORAGE

1. 行政办公室
2. 会议室
3. 楼梯
4. 入口
5. 卫生间区域
6. 残疾人卫生间
7. 淋浴
8. 卫生间
9. 技术设备
10. 样品室/静音室
11. 复印室/厨房/储藏室

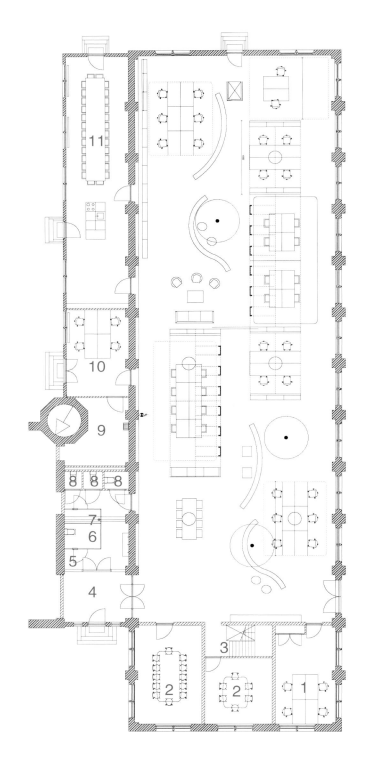

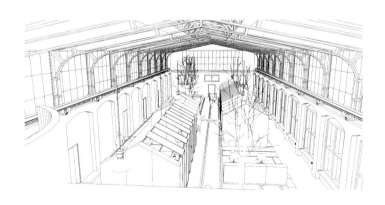

▶ DESIGN CONCEPT 设计理念

Fabrique's new offices house in an old industrial building of the Delft University. The size and industrial monumentality prompted the concept: MARKETHALL. The big space is filled with a few transparent 'market stalls' offering different atmosphere for different kinds of work.

　　Fabrique的新办公室位于代尔夫特大学一栋古老的工业化大楼内。规模及工业化的巨大外形促成了其设计理念：市场。大型空间里安置了透明的"市场摊位"，为各种不同工作提供不一样的氛围。

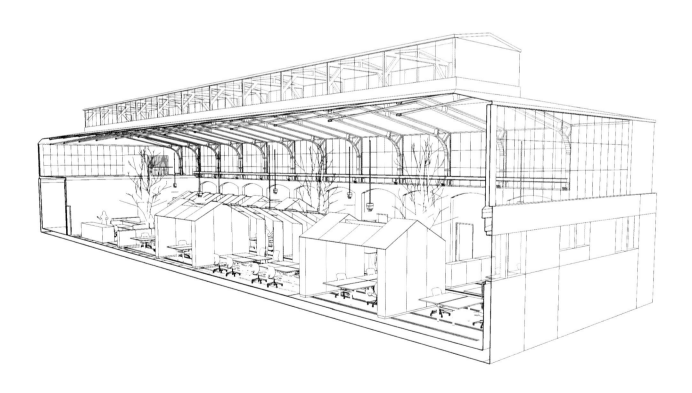

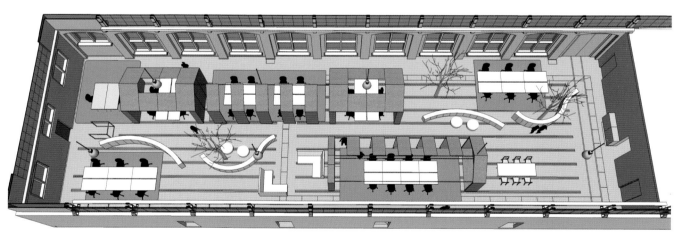
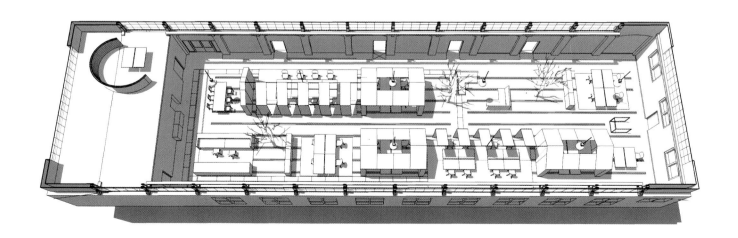

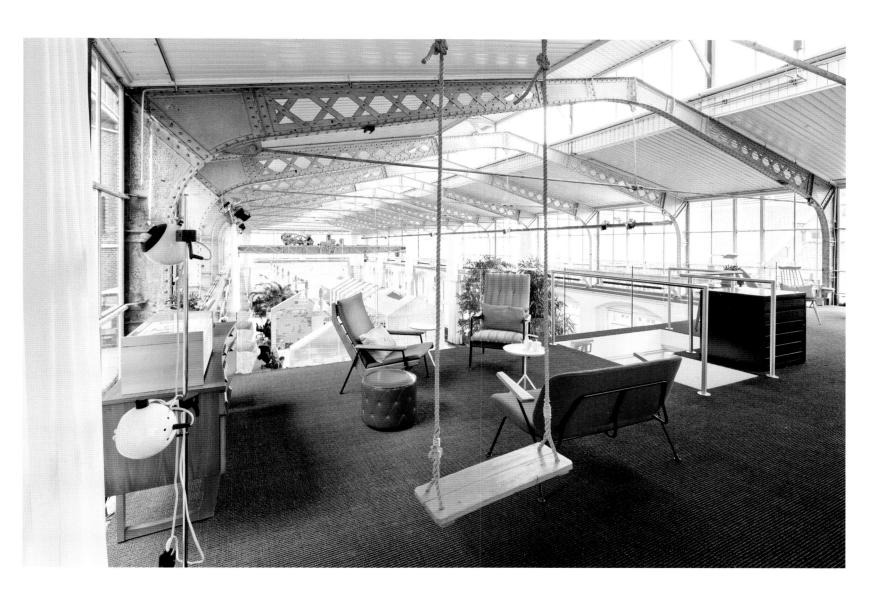

▶ FEATURE 特色

The existing building has a harsh industrial look & feel with lots of concrete, stone, glass and steel. Origins Architects wanted to add some lightness and softness so that to make it more natural and humane. Therefore they used natural materials (the wood in the OSB), soft materials (the felt, the molo-design room dividers) and green hues. The main hall has quite a lot of workspaces. Being a creative agency Origins Architects wanted to create an alternative playful space for Fabrique's employees that oversees the whole space. Above the hall, at one end there is a leisure area, here you can empty your mind for new ideas, have more relaxed meetings or just sit and read inspiring magazines. The market stalls are made of low-grade OSB panels.

At the other end, you'll see a yellow machine, which is part of the existing monumental building and was used to lift the heavy dynamos that were being assembled in this hall. Designers rolled it into a central position in the space and it was painted yellow. They believe it adds to an atmosphere of work and industry and keeps the original function of the building alive.

建筑的混凝土、石材、玻璃和钢材营造出粗糙的工业风观感。Origins建筑事务所希望增加一些亮度和柔软度，这样就会看起来更自然更人性化。因此，他们运用了自然材料（定向刨花板中的木材）、柔软材料（毛毡、莫罗设计感的房间隔板）以及绿色调。主大厅有很多办公空间。作为一家具有创意的企业，Origins建筑事务所希望打造一个既可供Fabrique员工们娱乐又可以纵观全局的空间。在大厅上方的一端有一个休闲区，你可以在这里放空自己、寻找新灵感，进行轻松的会议，或者只是静静地坐下来阅读能启迪人心的杂志。

在另一端，你可以看到一台黄色机器，它是现有纪念性建筑物的一部分，过去被用来升降安装在大厅内的沉重发电机。设计师将其转动到该空间的中心位置，并喷涂成黄色。他们认为这不仅增加了工作和工业化氛围，同时又保持了这栋建筑的原始功能。

Information Technology 信息技术

PROJECT LOCATION 项目地点	Singapore 新加坡	DESIGNERS 设计师	Penny Sloane, Junlong Lin 彭妮·斯隆、林俊龙	DESIGN COMPANY 设计公司	Siren Design Group Siren设计团队
PROJECT AREA 项目面积	3000 m²	PHOTOGRAPHER 摄影师	Owen Raggett 欧文·拉吉特		

Facebook
Facebook办公室

▶ CORPORATE CULTURE 企业文化

Facebook is an online social networking service headquartered in Menlo Park, California. Facebook is considered to be one of the fastest growing companies in the world. The culture at Facebook was open and transparent with no hierarchies. The company was well known for its 'hip geek culture' fostered by its founder Mark Zuckerberg. He tried to attract the best talent in the industry by creating a fun environment wherein employees had the opportunity to work on the best projects with a sense of openness.

Facebook是一家在线社交网络服务网站，其总部位于加州门洛帕克。Facebook被公认为世界上发展最迅速的公司之一。其文化是开放、透明、无阶层。该公司以其创始人马克·扎克伯格的"时尚极客文化"闻名遐迩。他试图通过打造一个风趣的环境来吸引最优秀的人才，在那里，员工有机会以公开、率真的感受在最好的项目中工作。

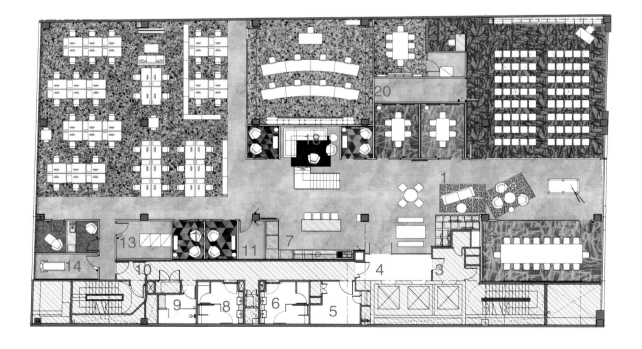

FLOOR PLAN 平面图

1. BREAKOUT LOUNGE
2. LARGE MEETING ROOM
3. FIRE FIGHTING LOBBY
4. LIFT LOBBY
5. STORE
6. M TOILET
7. MICRO KITCHEN
8. F TOILET
9. SHOWER
10. SMOKE STOP LOBBY
11. STORE
12. COSY
13. IT WIRING
14. TREADMILL MEETING
15. MASSAGE
16. GENERAL OFFICE
17. GAMING AREA
18. CHILL
19. CRISIS MANAGEMENT CENTRE
20. FURNITURE STORE
21. MEETING
22. GENERAL MEETING SPACE

1. 休息室
2. 大会议室
3. 消防大厅
4. 电梯间
5. 储藏室
6. 男洗手间
7. 迷你厨房
8. 女洗手间
9. 淋浴间
10. 无烟大厅
11. 储藏室
12. 温室
13. 信息技术配线室
14. 会议室
15. 按摩室
16. 总务处
17. 游戏区
18. 冷窖
19. 危机处理中心
20. 家具储藏室
21. 会议室
22. 全体会议区

▶ DESIGN CONCEPT 设计理念

Siren Design was appointed to design the 30,000sqft expansion of Facebook Singapore's office. Being their new AsiaPac HQ the design required the addition of Asia influences in the space whilst still having Facebook's brief of a raw industrial feel.

　　Siren设计公司受托设计Facebook3万平方英尺（2787m²）的新加坡办公室。作为新的亚太地区总部，本案需要在空间内加入亚洲影响力，同时保有Facebook原生态的工业化感觉。

▶ FEATURE 特色

One of the main features is a new industrial staircase connecting levels 14, 13 & 12. The staircase sets the stage for the collaborative vintage lounge breakout adjacent to the themed meeting rooms. To complete the design and global brand Siren Design commissioned three local artists to enhance the space with their interpretation of AsiaPac inspired graffiti. The space is colorful, warm and a pure joy to be in.

主要特色之一便是新建的连接12、13、14楼的工业化楼梯。该楼梯为主题会议室旁协同合作、古老的休闲室做足准备。为了完成该设计，全球品牌Siren设计公司委任三位当地画师，运用他们亚太灵感而成的涂鸦诠释来提升空间感。这个空间色彩鲜艳、温暖人心，是一个纯粹的快乐所在。

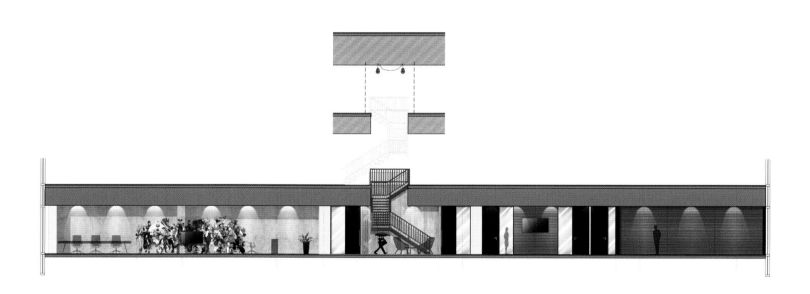

Information Technology 信息技术

PROJECT LOCATION 项目地点	Shanghai, China 中国上海	PROJECT LEADER 项目主管	John Yang 杨正茂	DESIGN COMPANY 设计公司	AECOM
PROJECT AREA 项目面积	2340 m²	PHOTOGRAPHER 摄影师	Zhong Chen 陈中	LEED TEAM LEED团队	Liang Liao, Peng Liu, Shiyuan Li 廖亮、刘鹏、李世元

Autodesk Shanghai Office
欧特克上海办公室

▶ CORPORATE CULTURE 企业文化

Autodesk is the leading 3D design software company in the world. Some of the company's well-known programs include Autocad, Revit, 3D MAD, MAYA and others, providing prominent digital design services for architecture, engineering and the media entertainment industry. The Autodesk Shanghai Office is located in the Pudong Lujiazui Software Park, and during the company's internal merger and expansion in 2012, Autodesk added two more floors in the C12 building also located in the park. This allowed for the accommodation of 200 employees.

Autodesk（欧特克有限公司）是全球领先的三维设计软件公司，其著名软件Autocad, Revit，3D MAX和MAYA等为建筑、工程以及传媒娱乐业提供卓越的数字化设计服务。Autodesk上海研发办公室位于浦东陆家嘴软件园，为适应内部整合及扩展需要，公司于2012年在园区C12楼新增2层办公楼面，可容纳约200名员工。

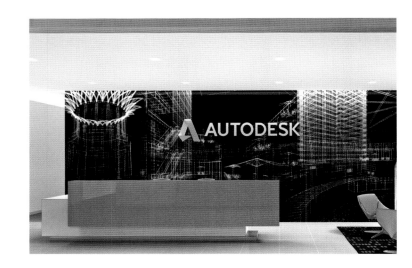

Sixth Floor (View from SW)
六层西南向鸟瞰图

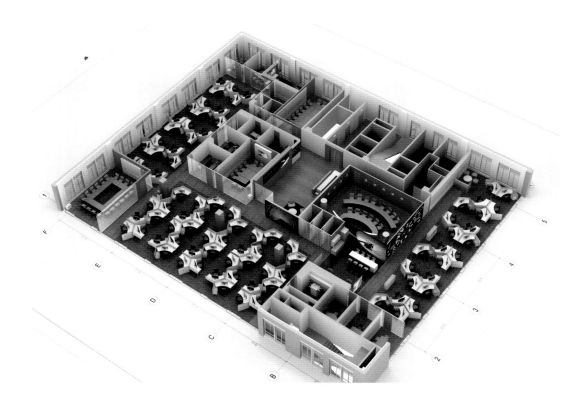

▶ DESIGN CONCEPT 设计理念

The objective of C12 project was to create a healthy and pleasant environment for the employees as well as to promote communication and cooperation among them. John Yang, design director at AECOM, created a creative design concept—a "mini city" full of vitality and passion. In this design concept, each department is within its own independent and quiet "street block," while "street corners" and "plazas" of varying sizes and shapes are spaces for informal teamwork and leisure. Also, walking through office corridors is not as simple as going from point A to point B. Instead, this "mini-city" design replaces mundane walks with urban strolls past "roadside" cafes and park benches. Employees will be lured out of their work spaces to occasionally walk along the streets and exchange information with their peers.

C12项目的设计旨在为员工打造一个健康愉悦的工作环境，亦能促进员工之间的交流协作。这让AECOM设计总监John Yang迸发出了一个奇思妙想的设计概念——将新办公室的布局设计为充满活力和激情的"迷你城市"，让部门与部门之间成为一个个相对独立、安静的"街区"，而不同大小和形状的"街角"和"广场"则成为非正式的团队合作区域和休闲场所。办公室走道变得不仅仅是从A点到B点那么简单，而是像"城市街道"一样，穿过"街旁"的咖啡座、"公园"的长凳再到多媒体室的台阶座椅，来吸引员工走出工作台，在此"不期而遇"进行多种形式、轻松的交流。

▶ FEATURE 特色

The Autodesk office gives a first impression of openness and transparency. In order to make maximum use of sunlight and create a well-lit working environment, open-concept working areas are placed close to the windows. Auxiliary functional spaces such as the manager's office, meeting rooms, print rooms, and telephone booths are located in the core area and separated by glass partitions. These spaces need to be both private and transparent at the same time, so the glass partitions are coated in two layers of matte film. Based on the software coding pattern, a mural representing the Shanghai skyline and street view was also created on the glass doors of meeting rooms. These patterns on the glass doors, along with the changing lighting, cast unique shadows and represent AECOM's philosophy of embedding local elements in their design, and their respect for local culture. The concept of "city blocks" is artistically showcased, while interior design adds more meaning from a perspective of history and culture.

开敞和透明是人们进入Autodesk办公空间的一个最大感受。为了最大化利用日照采光，给员工提供一个明亮的工作环境，AECOM室内建筑团队在平面规划中将开敞工作区布置在靠窗位置，而核心位置则设置经理办公室、会议室、打印室和电话亭等辅助功能空间，并用双层系统玻璃隔断分隔。为了达到透光和私密性要求，设计师在玻璃隔断上贴上双层磨砂贴膜，其中外侧的贴膜是由软件编码组成的上海标志性城市街景图案。玻璃贴图的连续性形成了富有光影变化的展示墙，充分体现出了一个国际公司对地方文化的尊重，且让"城市街区"得到具体和灵活的体现，室内空间的语言也增加了历史文化的内涵。

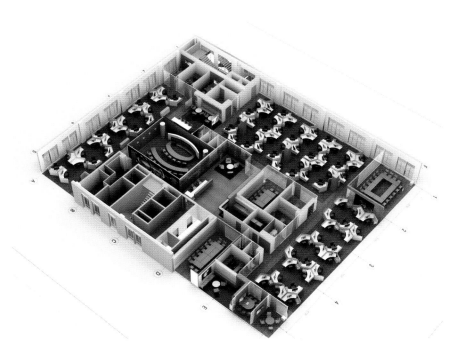

Sixth Floor (View from NE)
六层东北向鸟瞰图

Bright colors add energy to the office space, while 3D rendering wallpapers add an element of fun. These 3D rendering graphics are drawn with the 3D design software by Autodesk, representing the widespread use of Autodesk's products in various fields. Graphic design plays an important role in the design of the reception area. For example, a 10 meter by 3 meter Formica creative siding wall is imprinted with white 3D model line drawings of landmarks found in Shanghai.

鲜艳明亮的色彩为办公空间增加了能量，而生动活泼的3D渲染图墙面则为办公空间带来了不少趣味。这些3D渲染图案都是由Autodesk的三维设计软件绘制而成，展示了Autodesk产品在不同领域的广泛应用。图案设计同样在前台区域设计中扮演了很重要的角色，一副长10m、高3m的黑色Formica康贝特创意板墙上，印刻了白色的上海标志建筑3D模型线框图，作为一幅独特的展示背景，彰显Autodesk为建筑领域做出的卓越贡献。

The office also adopted Herman Miller's high performance workstation system which allows for 120 degree desktops to promote interaction and cooperation among employees. Low desktop partitions also create an open working space and a clean view. Both the proprietor and the employees unanimously agreed to sacrifice some personal workspace for larger communal areas, which, in turn promote communication and boost employee productivity. In addition, the walls of the meeting rooms and telephone booths are painted orange or yellow to contrast with the maple veneer furnishings. These details bring out a modern and vibrant working atmosphere for a high-tech company.

For interior and MEP design, AECOM uses Revit and 3D Max software developed by Autodesk. These programs allow for efficient communication between the designers, client and contractor, along for project to delivered and completed on time. AECOM blends Autodesk's corporate culture with the design, so its technique can be noticed everywhere in the entire office. This interaction creates a sense of participation for the proprietor and a sense of belonging for employees.

By using the 3D design software, we resolved issues involving geometry, color, lighting, interior design, and others. That allowed us to efficiently communicate with our clients and explore different design options. This approach accelerated the decision-making process and led to a design of high quality. —John Yang

开放办公区采用Herman Miller（赫曼·米勒）高性能工作站系统，120度灵活布局桌面，促进了员工之间的互动与协作；低桌面隔断保证了开敞办公空间的视线通透。适度牺牲些个人工作站的私密性和面积，相应增加员工交流的机会和公共活动区域，得到了业主和员工的共识。另外，大小会议室和电话亭都以橙色或黄色漆为背景，与枫木饰面系统会议桌形成生动的对比，体现了高科技公司优雅的现代文化和活跃的工作气氛。

AECOM在室内设计和机电设计过程中，全程使用Autodesk开发的Revit和3D Max三维设计软件，带给业主一个直观清晰的设计方案，让业主一目了然。正因如此，使得该项目的完成度十分之高。同时，AECOM又将Autodesk的产品及企业文化充分融入到了设计中，它的技术在其办公场所中无处不在。这种互动的形式，也让业主增添了一份参与感，增加了员工对公司的认同和归属感。

我们通过三维设计软件解决了诸如几何、色彩、灯光和室内设计等问题。这让我们与业主之间建立了有效的沟通和对探索不同设计方案的快速响应。通过这些方式加速决策过程来创造出一个高质量的设计。——杨正茂

Seventh Floor (View from SW)
七层西南向鸟瞰图

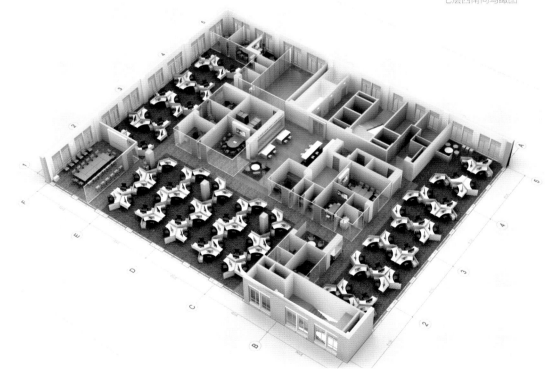

The design trend today is to build environmentally sustainable offices, and Autodesk requires all new offices to meet LEED criteria. AECOM's strength and experience in interior design ensured green and environmental protection factors in the Autodesk project. AECOM interior designers worked with its LEED team to implement multiple green design strategies. Some of these included strict control lighting and power density and water conservation measures. Intelligent lighting control equipment measures the levels of natural sunlight in order to provide a luminous interior environment. Other features include water-saving sanitary fittings and low-VOC finishing materials and furniture. 80% of the electronic equipment complies with Energy Star standards. During the process of construction, AECOM helped the main contractor manage green construction. AECOM assisted in developing a construction management plan, having special construction training and guideline before the main contractor commences work, giving assistance in executing air quality management during the process, and guiding the test of interior air quality before moving in. This project won the gold award certification of LEED-CI and had achieved the design objectives of both environmental protection and energy saving as well as health and comfort.

环保、可持续性设计是当今办公室设计的趋势，也是Autodesk对每一处新建、改建办公室的设计要求。而这一要求正好体现了AECOM在室内建筑项目中的综合实力，AECOM LEED团队的丰富经验也确保了此项目将绿色环保贯彻始终。AECOM室内建筑团队在设计中与LEED团队密切合作，采用了多种绿色设计策略，其中包括：进行采光分析，作为平面布置准则，对照明功率密度的严格控制，安装灯光系统的智能化控制装置，实现了办公区的自然采光，以此提供良好的室内光环境；使用节水型洁具以及低-VOC的装修材料和家具，并做到80%的电子设备达到Energy Star的标准；在施工过程中，AECOM协助施工总包进行绿色施工管理，包括协助制定施工管理计划，在施工总包进场前进行专门的施工培训和指导，施工过程中协助制定并执行空气质量管理计划，入驻前对室内空气质量检测进行指导等。该项目已经通过LEED-CI金奖认证，达到兼顾环保节能和健康舒适的设计目标。

AECOM definitely achieved the project's sustainability goals while creating an office that is also very beautiful.—Jenny Lum, Senior Program Manager for Autodesk.

AECOM达到了项目的可持续性设计的目标，同时又创造了一个优美漂亮的办公室。——Jenny Lum, Autodesk资深项目经理

Media
传媒

▲ If you have a dream-creating space
如果您拥有一个创造梦想的空间

▲ You can decorate it as this
可以这样装扮它

▲ Or in this way
也可以这样装扮它

▲ And also in another way
还可以这样装扮它

▲ Here, you can always find your dreamy office space
在这里，您总能找到属于自己梦想的办公空间

Media 传媒

PROJECT LOCATION 项目地点	Cape Town, South Africa 南非开普敦	DESIGNERS 设计师	Aidan Hart, Jenine Bruce 艾丹·哈特、詹宁·布鲁斯	DESIGN COMPANY 设计公司	Inhouse Brand Architects Inhouse Brand建筑事务所
PROJECT AREA 项目面积	850 m²	PHOTOGRAPHER 摄影师	Riaan West 里安·韦斯特		

John Brown Cape Town Offices
约翰·布朗媒体公司开普敦办事处

▶ CORPORATE CULTURE 企业文化

John Brown Media has a fast-paced, creatively-fueled office culture that is highly professional, yet not extremely formal. Meetings with clients in the lifestyle industry happen on a regular basis on the premise. This ensures for an ongoing vibrant and integrated atmosphere. Inhouse's design directly caters to their need for stimulating informal meeting spaces and a dynamic office environment which encourages quality creative output.

约翰·布朗媒体公司拥有快节奏、创意十足的办公室文化，非常专业却又不过于正式。在这个工作场所，会定期举行与生活方式产业领域客户的会议。这确保了持续、有活力、协调的气氛。Inhouse的设计直接满足了他们的需求，即拥有激励性的非正式会议空间以及可以鼓励高品质、创新型作品的充满活力的办公环境。

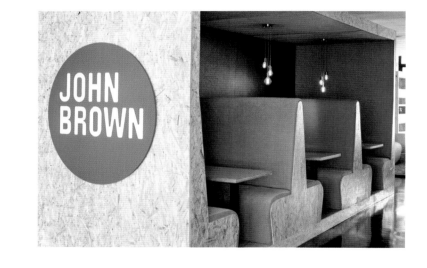

FLOOR PLAN 平面图

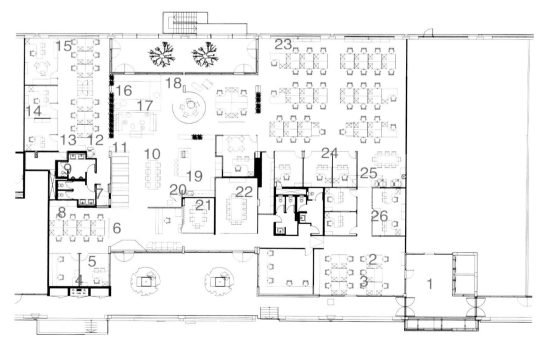

1. LIFT LOBBY — 1. 电梯大堂
2. PRODUCTION — 2. 成品室
3. FUTURE BUSINESS — 3. 未来商贸室
4. EXISTING TALL OPEN SHELF & CUPBOARD — 4. 高开架式书架及橱柜
5. DISCOVER OFFICE — 5. "探索"办公室
6. RECEPTION — 6. 接待室
7. SHOWER — 7. 淋浴室
8. EXISTING TALL CUPBDARD — 8. 高橱柜
9. WC — 9. 洗手间
10. CAMPSITE AREA — 10. 营地区
11. PAPER TABLE — 11. 文件桌
12. PRINT AND LAYOUT STATION — 12. 打印设计区
13. EXISTING CUTTING TABLE — 13. 裁剪桌
14. EXISTING LARGE BOOKSHELF — 14. 大书架
15. FINANCE — 15. 财务室
16. RAISED PICNIC AREA — 16. 上升就餐区
17. BEANBAG — 17. 懒人沙发
18. CASUAL SEATING AREA — 18. 休闲落座区
19. KITCHENETTE — 19. 小厨房
20. KITCHEN UNIT — 20. 厨房用具
21. MEETING ROOM — 21. 会议室
22. BOARDROOM — 22. 会议室
23. EXISTING SMALL BOOKSHELF STORAGE — 23. 矮书架储藏室
24. EDGARS OFFICE — 24. 埃德加办公室
25. PRINT AND LAYOUT STATION — 25. 打印设计区
26. EXISTING TALL OPEN SHELF& CUPBOARD — 26. 高开架式书架及橱柜

▶ FEATURE 特色

Adjoining the office's campsite area is an unconventional meeting room that takes the form of an oversized weaver's nest. The 'nest' is constructed from timber fins and is clad in strips of pine. It takes centre-stage in reference to John Brown's fondness for birds.
According to Inhouse designer Jenine Bruce:
"The nest was influenced by the company's incorporation of birds within their branding. So, we chose to use this in our design, building a nest which creates a 'safe' and intimate meeting spot for employees."
The existing large glazing facade allows natural light to filter in through the slats of the nest giving the interior a dreamy ambience. The nest has great presence and can be seen, via the keyhole design in the reception wall, as one enters the offices. The nest and lounge were set back against floor-to-ceiling glass walls that lead out onto a balcony and were positioned in such a way so as to take full advantage of the impressive views of Devil's Peak and Table Mountain.

与营地区毗邻的是一间非传统型会议室，做成了大号编织鸟巢的造型。"鸟巢"根据原木鱼鳍而构思，以松木条裹覆。这占据了重要位置，符合约翰·布朗对于鸟类的喜爱。

根据Inhouse的设计师詹宁·布鲁斯所述：

"公司的品牌中融入了鸟类的概念，鸟巢便是受此影响。所以，我们选择在设计中使用它，为员工打造一个'安全'、亲密的会议地点。"

现有大型玻璃立面使自然光线通过鸟巢的缝隙穿透进来，带给室内空间一种梦幻的氛围。鸟巢非常大型，人们进入办公室时，透过接待墙的孔型设计即可看到。鸟巢和休息室背靠落地玻璃幕墙，朝向阳台，这一有利位置让人们可以尽情享受魔鬼峰和平顶山带给人们的难忘景观。

DESIGN CONCEPT 设计理念

The re-design of John Brown Media's premises aims to increase connectivity among staff members while aligning the physical environment with John Brown's charged work ethic and creative output.

In order to achieve this, Inhouse created various inviting informal meeting spaces. The offices open into a shared space that includes an open-plan kitchen with a breakaway dining table and seating. Directly opposite the kitchen and table is a series of three eating pods featuring high-back booth. Because of its proximity to the kitchen, the booths have successfully enabled staff to socialise and engage with one another while grabbing a bite to eat.

Beyond the kitchen and dining area lies a lounge area or indoor "campsite". The "campsite" is kitted out with faux grass, picnic-like seating, timber plant boxes and canary yellow wooden tree-trunks. Brightly hued lounge seating, a cartoon-covered 'bed' and birdcage light fixtures add delightful visual fodder. Multi-functional, the campsite serves as a versatile informal lounge, communal gathering space, and waiting area for guests. This "campsite" area allows staff to revitalise, and has catalyzed a creative, happy working environment.

Flanking these main feature areas on either side lie open-plan offices for staff, with the central campsite acting as a visual and physical connector.

The staff workstations are partly screened off by the series of multi-functional wooden shelving that supports both planting and library space and provides some measure of privacy.

约翰·布朗媒体公司再次设计的前提，旨在提高员工之间的连通性，同时使物理环境与约翰·布朗紧张的工作氛围与创造性成果相一致。

为实现这一目标，Inhouse创建了各种诱人的非正式会议空间。办公室通向一个共享空间，包括带有分离式餐桌椅的开放式厨房。正对着厨房和餐桌的是三个用餐空间，以高背餐位为特色。因为邻近厨房，这些餐位使得员工可以一边简单吃点东西，一边进行社交。

离厨房和用餐区较远的那边是休息区，或称为室内"营地"。这个"营地"装备着人造草坪、野餐式座位、木料箱子以及淡黄色木质树干。明亮色调的躺椅、卡通表面的"床"和鸟笼状灯架增添了令人愉悦的视觉素材。该营地具有多功能性，可作为通用非正式休息区、公共聚集地或者来宾等候区。这个"营地"让员工活力再现，催化了一个创造性的、欢乐的工作环境。

这些主要区域的侧面设有开放式员工办公室，中央的营地扮演着视觉与物理层面联结的角色。员工的工位有一部分被多功能的木质架子隔挡，既支撑植物及图书空间，又提供某种程度的私密性。

Media 传媒

PROJECT LOCATION 项目地点	Moscow, Russia 俄罗斯莫斯科	CHIEF ARCHITECT 首席建筑师	Dmitry Ovcharov 德米特里·奥夫沙罗夫	ARCHITECTURAL COMPANY 建筑公司	Nefa Architects Nefa建筑工作室	CHIEF ENGINEER 总工程师	Sergey Kurepin 谢尔盖·库瑞皮	PHOTOGRAPHER 摄影师	Alexey Knyazev 阿列克谢·克尼亚杰夫
PROJECT AREA 项目面积	8800 m²	AUTHORS TEAM 顾问团队	Dmitry Ovcharov, Maria Yasko 德米特里·奥夫沙罗夫、玛丽亚·雅思司	GENERAL CONTRACTOR 承建商	ATITOKA ATITOKA建筑公司	MANAGEMENT COMPANY 管理公司	Cushman & Wakefield 库什婴&韦克菲尔德		
ARCHITECTS 建筑师	Victor Kolupaev, Olga Ivleva 维克多·克鲁佩伍、奥尔加·爱勒夫	LIGHTING ENGINEERS 灯光工程师	Spector Lab, Moscow, Russia 俄罗斯莫斯科斯佩克特工作室	PROJECT MANAGEMENT 项目管理	Daria Turkina, Maria Boyko 达莉亚·图尔基娜、玛丽亚·博伊科				

Leo Burnett
里奥·贝纳广告公司

▶ CORPORATE CULTURE 企业文化

Leo Burnett Company, Inc., otherwise known as Leo Burnett Worldwide, Inc., is an American globally active advertising company, founded in 1935 in Chicago by Leo Burnett. Their philosophy, Creativity has the power to transform human behavior, is the core belief of what they call HumanKind. They strive to cultivate an environment that celebrates creativity in all its forms, and that challenges people to grow and develop their talents in a supportive community. As a part of Leo Burnett Worldwide, Inc., this specially designed office is located in Russia.

李奥·贝纳股份有限公司，也被称为李奥·贝纳国际股份有限公司，1935年由李奥·贝纳创建于芝加哥，是美国一家活跃的全球化广告公司。他们的经营理念——创意拥有改变人类行为模式的力量，是他们所谓人类的核心信念。他们力求培养一种能够以任何形式激发创造性的环境，使人们在一个支持性的团体内接受成长与开发潜能的挑战。这间特别设计的办公室是李奥·贝纳国际股份有限公司的一部分，坐落于俄罗斯。

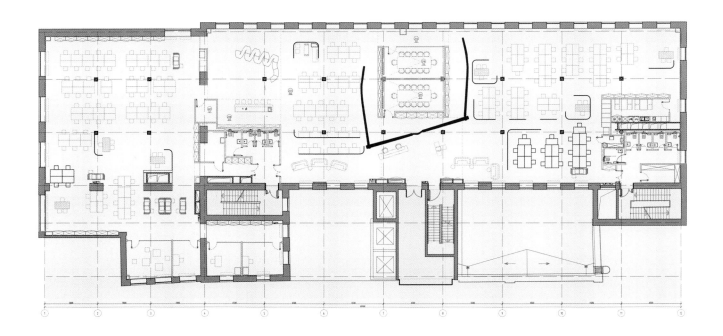

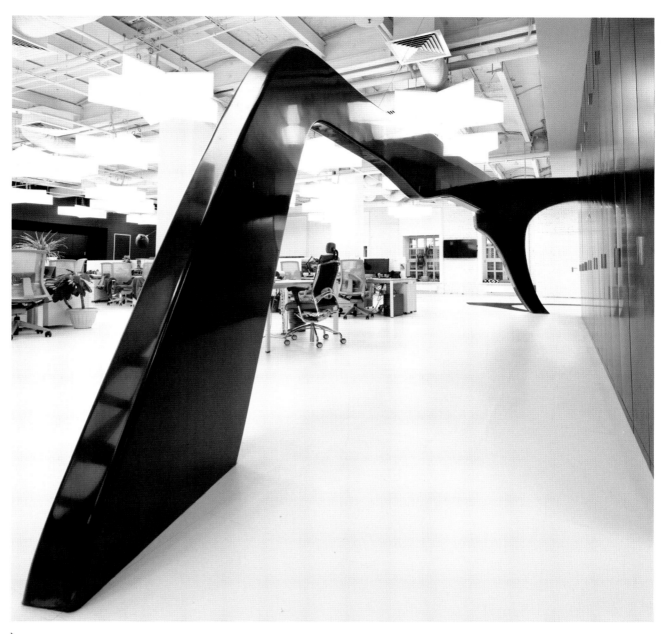

▶ DESIGN CONCEPT 设计理念

"When people are creative, work can be a bit chaotic. We've created a chaos that's nevertheless organized," designer Maria Yasko says. "We tried to save the loft feeling," Ovcharov says. He and Yasko certainly took advantage of the impeccable industrial credentials. The huge arched windows, brick walls, high ceilings, and exposed ductwork at Leo Burnett now complement a black-and-white overall scheme, which lends an extra jolt to the red of corkscrew-shape lounge seating and oversize task lamps.

"We gave Leo Burnett office a gimmick as a kind of compositional center —the glasses," Yasko says. Scribbled, seemingly, on their sides are Burnett's creative mantras. For example, "Creativity has the power to transform human behavior."—Quoted from Interior Design Magzine

设计师玛利亚说，"当人们富于创造力的时候，工作会有一点凌乱。我们创建了一种有条理的混沌。"奥夫沙罗夫说，"我们试图保留阁楼的感觉。"他们两人充分利用了无可挑剔的工业化资质。李奥·贝纳巨大的拱形窗、砖墙、高高的天花，还有外露的通风管道整体搭配了黑白的色彩方案，而红色的螺旋状躺椅及特大号工作灯具增添了额外的跃动感。

玛利亚说："我们赋予李奥·贝纳办公室一种巧妙的创造中心——玻璃。"上面有着涂鸦式的李奥·贝纳的创意颂歌。比如，"创意拥有改变人类行为模式的力量。"——引自《室内设计》杂志

▶ FEATURE 特色

"The idea to use glasses as the central expressive feature of this space was not born at once," said designer Maria Yasko. "We considered stars and apples, which are central to the company's signature style."

"The first agency was established in Chicago in 1935, and we have aspired for some style reference to that period. When we brought glasses scaled up several times and shaped like those worn by Leo Burnett, we thought it looked cool, and even more so if we would hang big glittering stars to the ceiling."

"This is not the first time we have designed working spaces for ad agencies," Dmitry Ovcharov added. "Leo Burnett's previous office had meeting spaces and rooms located inside giant pencils."

"Developing the concept for Publicis Group this time, we have opted for an idea of office spaces as modern art galleries. We did not want to make something complex; the glasses are integrated with the space, serving as a purely artistic feature and exposition, but on the other hand, being exactly the solution that organizes the space."

"The interior may or may not get some additional drawings or slogans or a new color in the future. That is the idea. The environment may change, but what will always be maintained is the core expressive substance, which is self-sufficient and makes the space complete."

"用玻璃作为本空间居中表现特色的想法由来已久，"设计师玛利亚说，"我们考虑了星星和苹果，它们对这间公司的标志性风格都很重要。"

"1935年，第一间机构成立于芝加哥，我们向往跟那个时期有关的风格。当我们把玻璃按比例放大几次，做成里奥·贝纳穿戴的那样，我们认为它看起来很酷，而且如果我们将大型的金光闪闪的星星挂在天花板上效果会更好。"

"这不是我们第一次为广告公司设计办公空间，"德米特里说，"里奥·贝纳以前的办公室在巨大的铅笔中有会议场所和房间。"

"这次为阳狮集团发展理念，我们选择现代画廊般办公空间的灵感。我们不想做复杂的东西，玻璃融入空间，作为纯粹的艺术特色及展示，但另一方面，也恰好是构成空间的解决方案。"

"在未来，室内可能会加入一些额外的挂画、标语或者新色彩，也可能不会。这仅仅是个构想。环境或会改变，但是自给自足的核心表现实质永远不变，它使空间变得完整。"

Media 传媒

PROJECT LOCATION 项目地点	Stockholm, Sweden 瑞典斯德哥尔摩	DESIGNER 设计师	Peter Sahlin 彼得·萨林	ASSISTING ARCHITECT 助理设计师	Mikael Hassel 米卡埃尔·哈塞尔	LIGHTING 照明设计	Beata Denton 贝亚特·丹顿	DESIGN COMPANY 设计公司	pS Arkitektur pS建筑公司
PROJECT AREA 项目面积	4000 m²	PHOTOGRAPHER 摄影师	Jason Strong Photography 杰森·斯特朗摄影	ARCHITECTS 建筑师	Beata Denton, Therese Svalling (interior design), Peter Sahlin, Leif Johansen (Building Design) 贝亚特·丹顿、泰雷兹·斯瓦林（室内设计）、彼得·萨林、列夫·约翰森（建筑设计）				

Chimney Group
Chimney集团办公室

▶ CORPORATE CULTURE 企业文化

Chimney Studios are storytellers at heart and use film and communication expertise to produce over 6,000 outputs for more than 60 countries every year. The work covers each step of the creation process, from conception through development, production and world-class post-production. Areas include feature films, commercials and TV. Founded in Stockholm in 1995 and now engaging 300 employees in offices in 10 countries. The Stockholm office employs some 190 people.

　　Chimney集团工作室本质上是故事的讲述者，每年运用电影和交流专业知识为60多个国家输出6000多部作品。他们的工作涵盖创作过程的每一步，从概念到发展，从成品到世界级水平的后期。其领域包括故事片、商业广告和电视剧。该公司于1995年在斯德哥尔摩成立，现今在10个国家拥有300名员工。斯德哥尔摩办公室大概有190名。

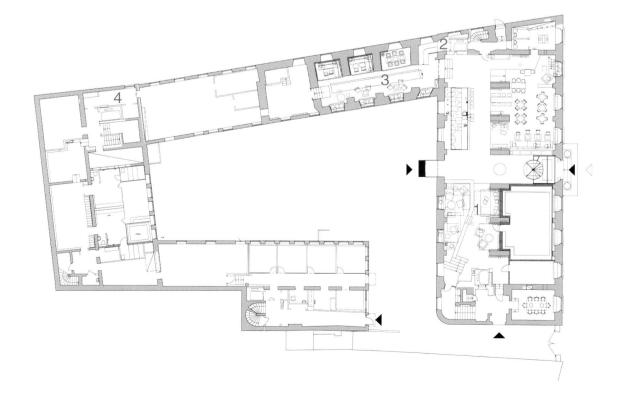

FLOOR PLAN 平面图

1. PENDANT METAL CEILING
2. CABINET FOR ELECTRONICS
3. CARPET RUNNER
4. CABINET FOR ELECTRONICS

1. 金属顶饰天花板
2. 电子设备橱柜
3. 通道地毯
4. 电子设备橱柜

▶ DESIGN CONCEPT 设计理念

The challenge was to fit this friendly, open minded, top-of-the art technical and busy company into an 18th century classified building on Stockholm´s most picturesque walkway by the waterfront.

The entrance level houses a great lobby under the vaults. This is where you meet colleagues, have lunch or work over a cup of coffee. Foreign visitors from the film industry enjoy the cool atmosphere and the professionalism from staff and studios.

The design concept is The Chimney Family—welcoming, professional and sporting gifts and souvenirs from friends all over the world. The interior is a mix of second hand and new furniture, all held together by a toned down color scheme and soft fabrics.

将这个友好、开放、艺术科技领域首屈一指且业务繁忙的公司融入斯德哥尔摩最别致的水滨通道上18世纪的建筑，是一项挑战。

入口层面覆盖了一个有拱顶的大型休息室。在这里，你可以遇见同事，一起享用午餐或者边工作边喝杯咖啡。电影界的外国客户非常欣赏这种酷酷的氛围以及员工和工作室的职业化。

本案的设计理念是Chimney集团大家庭来自世界各地的朋友馈赠的热情、专业、公正的礼物和纪念品。室内空间是二手家具和新家具的混搭，通过一种柔和的色彩方案及柔软的布料整合在一起。

▶ FEATURE 特色

Chimney Studios occupy most of the 7 stories of the building. Basically no alterations on existing structures were allowed. Light fittings and acoustic panels had to be fitted so as not to make any permanent impact. Besides office spaces it houses a first class cinema, sound studios, editing suites, grading suites and mixing studios. All of these are totally in-situ built light- and sound insulated units.

Chimney集团工作室占据了这栋七层楼的绝大部分。现有框架结构基本上没有改变。灯具配件及隔音板需要匹配，免于引起任何永久性冲击。除了办公空间，这里还包括一间头等影院、声音工作室、剪辑室、定级室和混音室。所有房间都在现场建造，隔绝光源和声音。

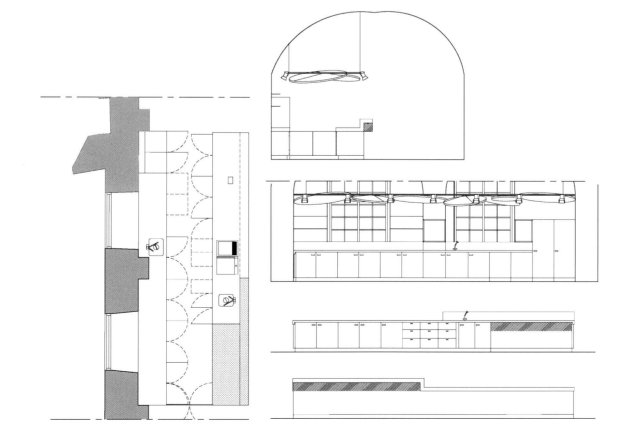

PROJECT LOCATION 项目地点	Istanbul, Turkey 土耳其伊斯坦布尔	DESIGN TEAM 设计团队	Selçuk Avcı (Founder / Head Architect), Deniz Nar (Partner), Burak Ünder, Buşra Al, Tuğba Öztürk 塞尔丘克·阿夫西（创始人及首席建筑师）、丹尼斯·纳尔（合伙人）、布拉克·安德、鲍撒、阿尔、坦布·欧兹克
PROJECT AREA 项目面积	1450 m²	PHOTOGRAPHER 摄影师	Asiana Jurca Avci 阿斯那·朱卡·阿夫西	DESIGN COMPANY 设计公司	Avci Architects Avci建筑公司

FOX International Channels
福克斯国际频道办公室

▶ CORPORATE CULTURE 企业文化

FOX International Channels is 21st Century FOX's international multi-media business. They develop, produce and distribute 300+ wholly- and majority-owned entertainment, sports, factual and movie channels in 45 languages across Latin America, Europe, Asia and Africa.

福克斯国际频道是21世纪福克斯公司旗下的一家国际多媒体企业。它以45种语言进行开发、创作，并在拉丁美洲、欧洲、亚洲及非洲的300多个娱乐、体育、事实、电影频道拥有全部或大部分股权。

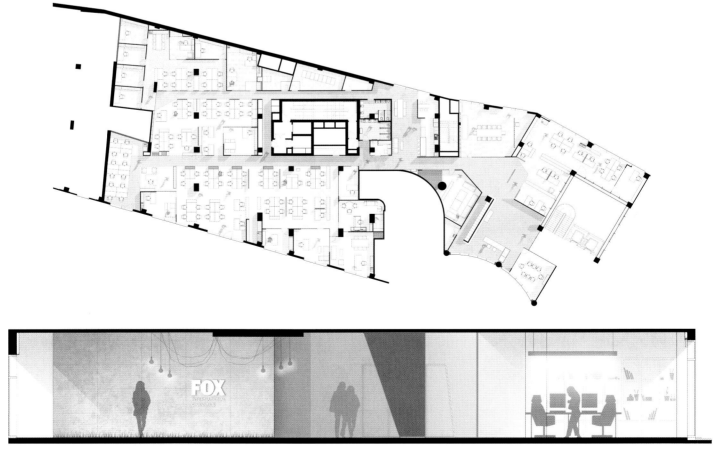

▶ DESIGN CONCEPT 设计理念

When Fox International Channels consulted Avci Architects for the design of their new office space, they described their dream working environment as a capacious, dynamic, vivacious and colorful space. A working environment possessing an open office layout has been created within the total area of the 1,450 sqm construction site. Besides the open plan offices, meeting rooms and closed plan offices with a glass structure in dynamical geometrical forms were projected.

After three months of the design process, the construction was completed in two months through which the use of natural material was the determinative element in its design on the grounds of common precision shared both by Fox International and Avci Architects. The use of natural light was maximized by putting the transparency to the forefront in the division of offices. In this office, which was designed with an aim to make the long hours of work more comfortable, the use of vivid colors and graphics on the walls intend to make the working environment rich and energetic without tiring the employees.

当福克斯国际频道向Avci建筑公司咨询新办公室设计的时候，他们把梦想的工作环境描述为一个宽敞、动态、有生气、色彩缤纷的空间。在总面积1450m²的工地，打造了一个拥有开放式办公布局的工作环境。除了开放式办公室，预期还将构建带有动态几何图案玻璃结构的会议室及封闭办公室。

在设计过程三个月后，该建筑于两个月内竣工，自然材料的运用在地面设计中是决定性因素，其精度由福克斯国际频道及Avci建筑公司共同确认。通过将办公室区隔的透明度推到最前端，自然光线的运用得到最大化。在这间办公室内，设计意图是让长时间的工作变得舒适，墙壁上生动色彩和图形的运用使工作环境丰富、充满活力，而不使员工感到疲累。

▶ FEATURE 特色

When tackling the setup of the office plan, first of all the layout of the departments and common areas were assessed. The initial idea for the design was to allow a creative and free working environment through providing spaces appropriate for recreative and free use, in line with the vibrant structure of Fox International Channels.

在处理该办公室规划设置的时候，首先评估了所有部门和普通区域的布局。设计的初衷是通过提供适合消遣、免费使用的空间建成一个创意十足、自由的工作环境，符合福克斯国际频道充满活力的构造。

Media 传媒

| PROJECT LOCATION 项目地点 | Moscow, Russia 俄罗斯莫斯科 | GRAPHIC DESIGN 平面设计 | Yulia Dorokhina 尤利娅·多罗基娜 | ARCHITECTURAL SUPERVISION 建设监理 | Olga Ivleva, Sergey Kurepin 奥尔加·艾维莱瓦、谢尔盖·库雷宾 | ARCHITECTURAL STUDIO 建筑公司 | VOX Architects VOX建筑公司 |
| PROJECT AREA 项目面积 | 6500 m² | PHOTOGRAPHER 摄影师 | Alexey Knyazev 阿列克谢·克尼亚杰夫 | GENERAL CONSTRUCTOR 主施工单位 | Atitoka | PROJECT MANAGEMENT 项目管理 | Cushman & Wakefield, Ekaterina Afanasieva, Tatiana Shumulikhina 库什登&威克菲尔德、叶卡捷琳娜·阿法娜锡耶瓦、塔蒂阿娜·叙穆莉齐那 |

Publicis Groupe
阳狮集团办公室

▶ CORPORATE CULTURE 企业文化

PUBLICIS GROUPE is a French transnational advertising-communications holding. The third biggest in the world and the biggest one in Europe. It works in Russia since 1990 and is represented by media-groups: Starcom MediaVest Group Russia (Starcom and MediaVest agencies), ZenithOptimedia Group Russia (Zenith and Optimedia agencies), creative agencies Leo Burnett, Publicis, Saatchi&Saatchi, etc.

法国阳狮是一间跨国广告及传播控股集团，排名世界第三，欧洲第一。它自1990年起在俄罗斯运作，以星传媒体集团、实力传播集团、创意机构李奥·贝纳广告公司、阳狮广告、盛世长城等媒体公司为代表。

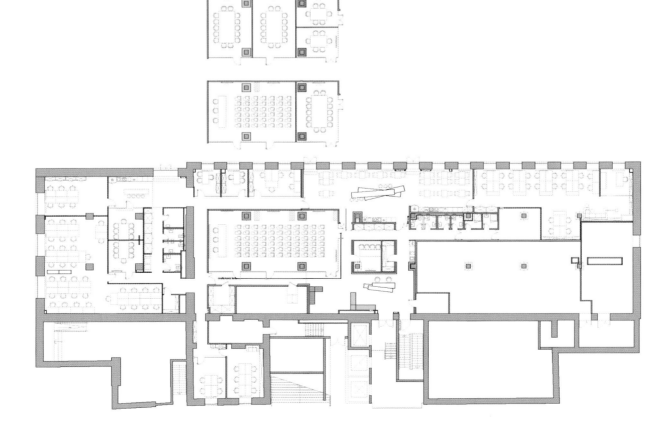

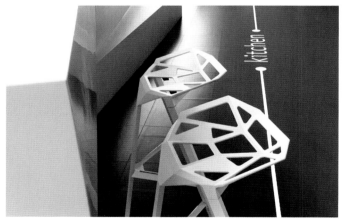

▶ DESIGN CONCEPT 设计理念

Boris Voskoboynikov team—VOX Architects—is experienced in creating interiors for this type of companies. In this particular case, when developing the project for PUBLICIS GROUPE, architects had faced an interesting challenge. There was no expressive brand-book language, as it was in LEO BURNETT project—working with it, generally, is easy; There were no vivid intentions from SAATCHI & SAATCHI art-directors, that each time grant the architects freedom to create somewhat new, extraordinary.

The PUBLICIS GROUPE interior was to be created as neutral as possible: holding combines the agencies, that differ in the mood, brand identity and ideologically. In the same time, the challenge was forced by the condition to give the space capability to be personalized. The VOX Architects team found the laconic solution and created ART LIGHT COLOR concept.

VOX建筑公司旗下的鲍里斯·沃斯科博伊尼科夫团队在打造这种公司室内设计方面经验丰富。阳狮集团这个特殊的案子，让设计师们面临一个有趣的挑战。这里不像李奥·贝纳那个案子，没有表达商标的语汇，在设计过程中总体来讲很轻松。没有来自盛世长城艺术总监的明确意图，每次都给予设计师自由，打造一些新鲜、与众不同的东西。

阳狮集团的室内旨在越中性化越好：结合集团自身不同气氛、品牌标识和意识形态。同时，使空间容量个性化这一条件加剧了该项挑战。VOX建筑团队找到了简洁的解决方案，来打造艺术光色理念。

▶ FEATURE 特色

The main functional area of PUBLICIS GROUPE—is the unifying floor. This is the place, where the employees of different holding's divisions meet, hold the conferences, take part in negotiations; this area can be exploited by any company individually, as this is the place, where the largest transforming conference-hall is situated.

The COLOR. The main brand colors were identified from the brand-book of each holding's company. Brand colors were sublimated into the lines, which entangle the area, like threads, changing the material of the surfaces, uniting all PUBLICIS GROUPE brands in one space.

The LIGHT COLOR. Special lighting surfaces are used in the project, they form the structure, that unites the area. Lighting surfaces are managed by the special system that regulates the brightness of the light and tunes up any color. The space that has the white neutral light surface in customary mode is able to be changed dramatically due to the variation of light flow.

The ART. The authors were inspired by the creations of light artist James Turrell, famous for his experiments with color light. This time, the spectators and, moreover, participants, are the employees and guests of the holding.

阳狮集团主要功能区是一体化的楼层。在这个地方，不同控股的部门员工会面、举办会议、参加谈判；它可以由任何一家公司单独使用，因为这里坐落着最大的转换式会议大厅。

色彩。主要的品牌色彩通过每家控股公司的品牌图册得以确定。品牌色彩升华到线面之间，像纺线一样在该空间纠缠着，改变表面的材质，在一个空间之内结合了阳狮集团所有品牌。

光色。本案中运用了特殊的照明表面，形成了联合该区域的结构。照明表面通过特殊系统操控，调节灯光的亮度，转变为任意颜色。拥有惯例模式白色中性光表面的空间可以随着光流量戏剧化地变幻。

艺术。设计师的创作灵感源自灯光艺术家詹姆斯·特瑞尔的作品，他以其彩色光的实验而闻名。这一次，观众以及参与者，是控股公司的员工和宾客。

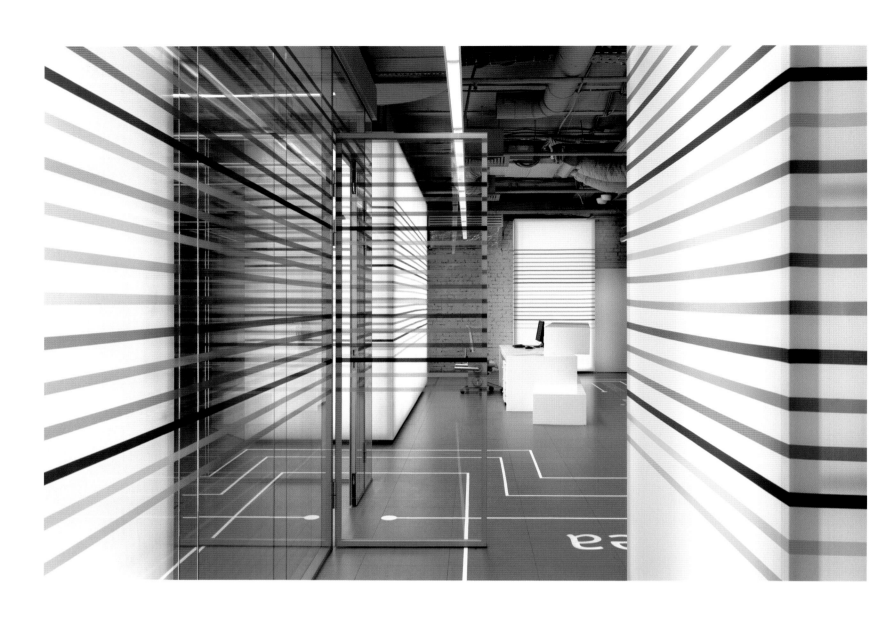
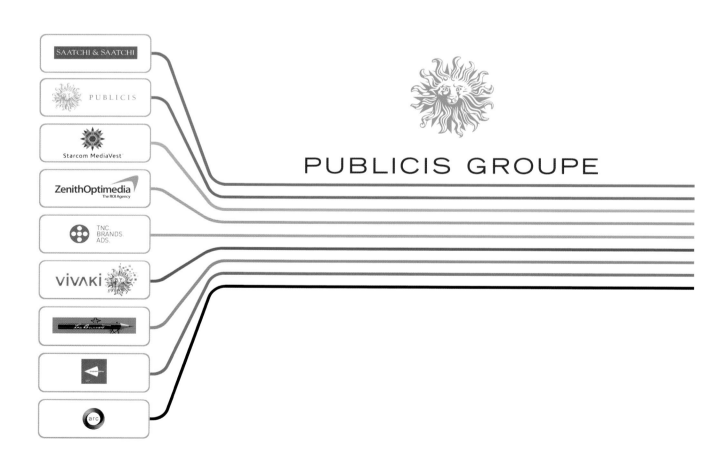

Design Consulting
设计咨询

▲ If you have a dream-creating space
如果您拥有一个创造梦想的空间

▲ You can decorate it as this
可以这样装扮它

▲ Or in this way
也可以这样装扮它

▲ And also in another way
还可以这样装扮它

▲ Here, you can always find your dreamy office space
在这里，您总能找到属于自己梦想的办公空间

Design Consulting
设计咨询

PROJECT LOCATION 项目地点	Moscow, Russia 俄罗斯莫斯科	DESIGN COMPANY 设计公司	IND Architects IND建筑事务所
PROJECT AREA 项目面积	280 m²	PHOTOGRAPHER 摄影师	Andrey Jitkov 安德烈·吉特考夫

Office of IND Architects Studio
IND建筑事务所办公室

▶ CORPORATE CULTURE 企业文化

Established in 2008, Ind Architects has principally practiced in architecture and the provision of architectural services. For them, love, philosophy and architecture are the same thing. They bring their love of life and work together in a rounded world view. They gently bring to people the advantages of natural light and happy, environmentally sound buildings with good values.

IND建筑事务所成立于2008年，主要从事建筑领域及建筑服务的提供。对他们来说，爱、人生观与建筑是一样的。他们以一种全面的世界观将对生活和工作的热爱结合在一起。他们逐渐将有着自然光线、让人愉悦且具有良好价值等优点的环保建筑物带给人们。

▶ DESIGN CONCEPT 设计理念

When working on the project of Ind Architects office, they entrusted themselves with the task to create a contemporary creative space with a minimum of distracting elements, so that designers and architects could focus on their projects as much as possible. The office has become a business card and a reflection of the studio which has implemented a number of projects over five years of its existence.

在IND建筑事务所办公室里工作时,他们交付自己一个任务,即运用最少的干扰元素打造一个现代化的创意空间,以便设计师和建筑师尽可能地集中精力完成项目。该办公室成为了其商业名片及工作室的映照,成立以来的五年间,他们已经实施了很多项目。

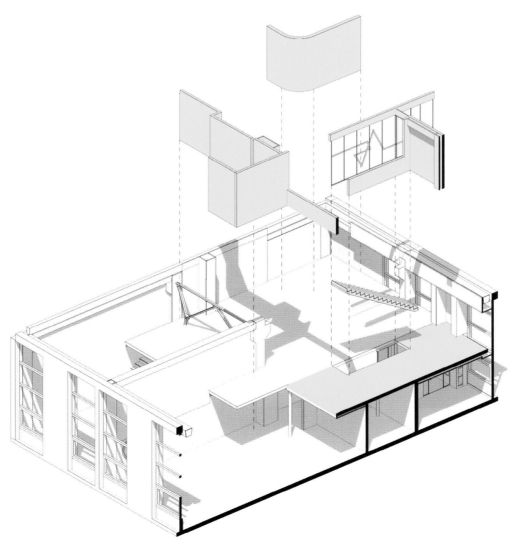

FEATURE 特色

The office has been arranged in ARTPLAY Design Center and has been decorated in the loft style. Designers have shown the benefits of a former industrial premise to their best advantage—they have kept the double-floor height area in some places, have deliberately left the concrete ceiling panels unpainted and have retained the original concrete structure—like timber shutter texture on the first floor and ceiling panels on the second floor. Exposed black-colored utilities accentuate the industrial past of the building and form a contrast to white walls of the office.

The office has been divided into two zones—one is a volumetric double-floor height area for designers and another one is a cozy space on the second floor for architects. The walls of the latter one are used for attaching the pictures and drawings of studio's current projects. In addition to two open space zones, a reception zone, a meeting room, and a coffee-point have been arranged on the first floor; an office, a leisure area, and an open meeting room for the staff may be found on the second floor. Two or three person meetings may also be held at a round table in the first-floor open space.

Grey and white are basic interior colors. Bright yellow details-such as infographics, a creatively different full-wall unicorn in the first-floor open space, and small items, like flower cache-pots, desk folders, and décor details-stand out sharply against quiet shades. The infographics had been developed as a part of a new corporate style of IND Architects and was patterned by the studio's architects.

In the leisure area, the employees may catch a break in a work process and talk about their goings, interesting projects and news while playing darts or foosball or read an interesting book nestling themselves down on a convenient pouf.

该办公室在ARTPLAY设计中心进行阁楼式装修。设计师将之前工业化场所的优点展示得淋漓尽致——他们在一些地方保留了两层的挑高区域，特意留下未喷涂的混凝土天花板，沿用原有的混凝土结构——比如一楼的原木百叶窗以及二楼的天花镶板。暴露的黑色设备凸显了这栋大楼的工业性，与办公室的白色墙面形成对比。

该办公室划分为两个区域——一个是两层挑高的设计师区域，另一个是二层舒适的建筑师区域。后者的墙壁上贴挂着工作室现有项目的图片和图纸。除了这两个开放空间，一楼还设有一个接待区、一间会议室和一个咖啡点；二楼则有一间办公室、一个休闲区和一间为员工准备的开放型会议室。两到三人的会议可以在一楼开放空间的圆桌旁进行。

室内以灰色和白色为主要基调。明亮的黄色则在细节上显现，例如在一楼开放空间的墙面上有着富有创造性的不同的独角兽图形，此外还有一些小物品，例如花盆、桌上的文件夹和其他装饰细节，它们在素净的色调下脱颖而出。信息图形已经发展成IND建筑事务所新型企业风格的一部分，并被工作室的建筑师仿照。

在休闲区，员工可以在工作过程中小憩，玩飞镖、桌上足球，或者悠闲地窝在便捷蒲团上阅读有趣书籍的时候，谈谈他们的工作情况、有趣案例及新闻等。

Design Consulting 设计咨询

ARCHITECTS 建筑师	Sergey Makhno, Ilya Tovstonog 谢尔盖·马克诺、伊利娅·托福斯通	PAINTING 绘画	Oleg Tistol 奥列格·提斯托尔		
PROJECT AREA 项目面积	200 m²	PHOTOGRAPHER 摄影师	Andrew Avdeenko 安德鲁·艾弗丹科	SCULPTURES, RELIEFS 雕刻、浮雕	Nazar Bilyk, Dmitry Greek, Sergei Red'ko, Yuri Musatov 纳扎尔·比利克、德米特里·格里克、塞奇·雷德科、尤里·穆萨托夫

Makhno Workshop
马克诺工作室

▶ CORPORATE CULTURE 企业文化

Makhno Workshop was founded by Sergey Makhno, who is a designer, architect and CEO of the studio. Their team takes care of every stage of a project—starting from a concept to its realization. They believe in traditions but do not limit themselves to the basics, trying to bring innovations, rich aesthetics, and functionality to any design solution. Crazy team of designers and architects visits office every day with energy and inspiration. Main rule—don't afraid experimenting!

马克诺工作室由谢尔盖·马克诺建立，他既是设计师、建筑师，又是该事务所的首席执行官。他们的团队专注项目的每个阶段——从概念出发，直到实现。他们相信传统，但并不把自己局限于基础领域，他们在任何设计方案中都努力带入创新、丰富的美学及功能性。疯狂的设计师及建筑师团队每天都带着能量与灵感来访办公室。主要规则是——不要害怕尝试！

▶ DESIGN CONCEPT 设计理念

Minimalism with loft elements and warm notes of Ukrainian art create a special atmosphere in the workshop that, like a mirror, reflects the owner's inner being. The key concept of the architect's office project is to implement the elements worked out by the designers to meet their customers' needs.

伴随着阁楼元素和乌克兰艺术的暖色调,极简派艺术在这间工作室里创造出了很特别的氛围,它就像一面镜子,反映出主人的内心世界。对于建筑师的办公环境设计的关键在于设计师制定出的方案要能和客户的需求相吻合。

▶ FEATURE 特色

The studio with a total area of 200 sq. m combines a showroom and a home-and-work place. "I had an idea to create a space that could inspire not only my team but also our clientele. We want our guests to take on a challenge facing brave and bold experiments", says Sergey. The studio with a total area of 200 sq. m combines a showroom and a home-and-work place. "I had an idea to create a space that could inspire not only my team but also our clientele. We want our guests to take on a challenge facing brave and bold experiments", says Sergey.

工作室的总面积为200m²，包括一个陈列室和一个办公休息区。谢尔盖说："我想创造一个空间，它不仅能激发我团队的设计灵感，还能带给我的客户启发。我们希望客户能够勇敢地接受挑战，大胆地尝试新颖的构思。"

When designing the studio Makhno used his favourite materials such as concrete, stone, glass, copper, bronze, various species of wood, high quality "TOTO" sanitary ware&fitting and "Miele" kitchen furniture. The office ceiling which is 4-metre high makes you feel easy just as in a free-open surroundings. The works of talented Ukrainian sculptors and artists create a special mood in the minimalist workshop. The bass-reliefs, unique statues, author's décor and furnishings are united to complete more than just an office but an original showroom of Ukrainian design and art. The main lighting ideas are conveyed based on Makhno's sketches.

The floor is covered with bulk concrete according to the laying technology which was applied for hospital rooms in 1930s. The flooring is considered to be ecologically pure and water-proof. Due to its smooth non-porous structure, it is void of cracks, joints and very resistant to bacteria.

在设计工作室的时候，马克诺用了他最喜欢的一些材料，比如混凝土、石头、玻璃、金属铜、青铜制品、各个品种的木材、高品质日本东陶制成的卫生器具、设备以及德国美诺公司的厨房用具。办公室的天花板距地面有4m高，这个高度会让你觉得就像在一个完全开放的环境里一样轻松。乌克兰颇有天赋的雕刻家和艺术家的作品给具有极简抽象派艺术特色的工作室营造了一种特殊的气氛。极具文艺气息的装饰品和家具的混合使用完成了一个远远超过办公室的环境，它更像乌克兰设计产品和艺术品的独特陈列室。主要照明设备的设计思路建立在马克诺画出的草图基础之上。

覆盖在地面上的大体积混凝土是根据20世纪30年代应用在医院房间里的敷设技术铺设的。地板的材料不仅是纯生态的而且还防水，由于它光滑无孔的结构，地面上既没有裂缝也没有接缝，甚至连十分顽固的细菌都没有。

The interior design was created by the studio owner with deep love and attention to his team and their common cause. Therefore, it has a function of an open home space with convenient comforts such as a cosy kitchen, bathroom facilities and a lounge zone. Sergey Makhno cherish the culture of Japan, the country of the rising sun, and honours a tea ceremony. For this reason, he designed a tea room in his studio where together with his friends and visitors the architect can enjoy sipping a fragrant drink. An important element in the workshop is a rich collection of Ukrainian zoomorphic ceramics that integrates harmoniously with the laconic concrete walls and bright Marc Newson "Felt chairs". Pottery in Ukraine has been developing from ancient times. The pieces of pottery art created by talented masters have spread worldwide. Ukraine's ceramic crafts reveal a history of the country's skills and talents. Makhno managed to pick out unique pottery pieces from different corners of Ukraine for them to take a place of honour in his private collection of ceramics.

工作室内部的设计产生于主人深沉的爱、他对团队的关心以及整个团队共同的目标。因此，它拥有开放的家庭空间便捷安逸的特点。舒适温馨的厨房、浴室设施和休息区都是这一特点的体现。谢尔盖·马克诺偏爱日本这个旭日东升般国家的文化，并且十分崇敬茶道。出于这个原因，谢尔盖在工作室里设计了一个茶室，设计师们可以和朋友或拜访者在这里享受啜饮香茗的休闲时光。工作室里存在着这样一个重要元素，里面拥有乌克兰动物形陶器的丰富馆藏，它们与简洁的混凝土墙壁及马克·纽森的鲜明"毡椅"和谐地共处一室。乌克兰的陶器制造术从古时候就开始发展。一件件由多才多艺的大师们制造的陶器传播到了世界各地。乌克兰的陶瓷工艺揭露了这个国家的技术和才能历史。马克诺把从乌克兰各个不同角落挑选出的独一无二的陶器成功地带回了事务所，这一件件琳琅满目的陶瓷制品就是他私人陶器收藏的荣誉象征。

When you enter the workshop the first thing you lay your eye upon is a huge copper reception wall, chandeliers and décor on Sergey Makhno's project and one of the main exhibits of the studio—the bronze statue "Rain" by Nazar Bilyk, the famous Ukrainian sculptor. Lighting, room temperature and sound are controlled with the help of smart technology system by using a corporate mobile phone. Entering through the three-meter high door you will find a spacious business venue for conducting important events and meetings. It is also decorated with the author's chandeliers and wooden panels. The room is furnished with Sergey Makhno glass table and Kristalia "Elephant chairs". The collection of Ukrainian ceramics decorates the library which counts more than 1000 books on design and architecture brought from all over the world.

当你踏进工作室的时候，首先映入你眼帘的是一堵宽阔的铜质接待墙、水晶吊灯和谢尔盖·马克诺设计项目里的装饰品以及事务所里主要的展览品之——乌克兰著名雕塑家Nazar Bilyk的青铜雕塑《雨》。照明设备、室内温度和声音都是在控制之中的，而且是受公用的移动电话作用在智能技术系统的帮助下进行控制的。从3m高的大门进入，你会看到一个专为指导重大事件和会议而设计的广阔商用空间。它同样是用充满文艺气息的水晶吊灯和木质嵌板装饰的。这个空间里放置着谢尔盖·马克诺设计的玻璃办公桌和Kristalia的"大象椅"。被乌克兰陶器装饰着的书柜上摆放着1000多本来自全世界的关于设计和建筑的书籍。

Design Consulting
设计咨询

PROJECT LOCATION 项目地点	Aker Brygge, Oslo, Norway 挪威奥斯陆阿凯尔布里格	DESIGNERS 设计师	Nikki Butenschøn, Anthony Williams 尼基·布坦恩斯肯、安东尼·威廉斯	DESIGN COMPANY 设计公司	Haptic Architects Haptic建筑公司
PROJECT AREA 项目面积	800 m²	PHOTOGRAPHER 摄影师	Inger Marie Grini 英厄·玛丽·格里尼		

Arkwright
Arkwright办公室

▶ CORPORATE CULTURE 企业文化

Established in 1987, Arkwright is a strategy advisory firm with a staff of about 80 professionals and partners, with offices in Hamburg, Oslo, Stockholm and Zürich. Arkwright is owned by senior staff.

At Arkwright, people are pragmatic about methodology and passionate about creating results for their clients. Because every client faces its own set of unique challenges, they believe that each one requires tailored, client-oriented advice based on strong analytical skills and deep business know-how.

Arkwright closely works with a wide range of private corporations—both small cap and large, blue-chip companies—all of which share the objective of optimizing their businesses and increasing profitability.

Arkwright是一家战略咨询公司，成立于1987年，拥有约80名专业人士及合伙人，其办公室分别位于德国汉堡、挪威奥斯陆、瑞典斯德哥尔摩以及瑞士苏黎世。Arkwright为企业高级职员所有。

在Arkwright，人们讲求实际的方法论，并且积极热情地为客户创造结果。因为每个客户都面临着独特的挑战，他们相信每个挑战都需要基于强大分析能力和深度商业技巧、以客户为导向的咨询进行量身定制。

Arkwright与各种不同型的民营企业紧密合作，不论公司规模是小是大，或者是蓝筹股，只要共享一个目标——优化业务、增加盈利。

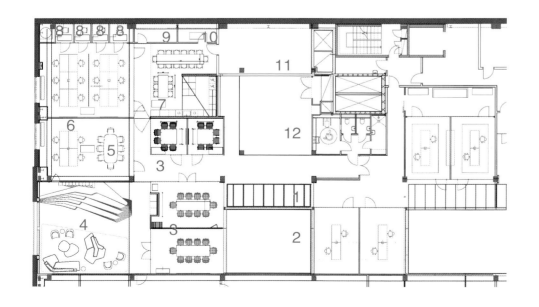

FLOOR PLAN 平面图

1. CORRIDOR	1. 走廊
2. O₂ ROOM	2. 氧气室
3. MEETING ROOM	3. 会议室
4. ASSEMBLY HALL	4. 会议厅
5. CONFERENCE ROOM	5. 会议室
6. OFFICE	6. 办公室
7. KITCHEN/DINING	7. 厨房/餐厅
8. QUIET ROOM/PHONE ROOM	8. 静音室/电话室
9. PRINT	9. 打印室
10. STORE	10. 储藏室
11. VENTILATION	11. 通风设备
12. JAMES BOND ROOM	12. "詹姆斯·邦德"室

▶ DESIGN CONCEPT 设计理念

The office is located in the prime harbour front location of Aker Brygge in Oslo, Norway, in an old converted warehouse building with a large arched window as its centrepiece. A new reception "sculpture" incorporates back offices, reception desk and a large stair/amphitheatre that straddles a double height space. The design is inspired by "svabergs", large granite stone formations that are typical for the area, rounded and polished by icebergs thousands of years ago.

该办公室位于挪威奥斯陆阿凯尔布里格主要海港的优越位置，在一栋古老、经过改装的仓储大楼内，大型圆弧窗作为装饰品在中心位置熠熠生辉。一个新的接待区"雕刻"、接待处、大型楼梯（跨越两层空间高度的阶梯室），组成了后勤办公室。该设计灵感源自"svabergs"，这是一种当地特有的大型花岗岩结构，几千年前经冰山的作用打磨而成。

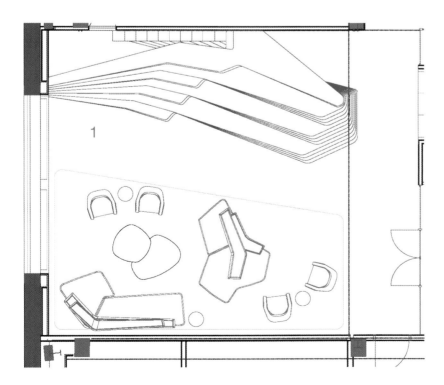

FLOOR PLAN 平面图

1. ASSEMBLY HALL   1. 会议厅

▶ FEATURE 特色

The new office space has been created for 40-50 employees, including workspaces, reception and back office, kitchen canteen, meeting rooms, breakout space and a "James Bond" room.
Special effort has been made to create a variety of spaces within the offices, incorporating green walls, double height spaces, and a special "James Bond" room.
The "James Bond" room is a windowless bunker-like space, sitting deep in the building—a difficult space to work with. This seemingly unpromising space has been transformed into an executive lounge, for quiet contemplation, creating a private, intimate and calming atmosphere.

　　新的办公室可容纳40~50名员工，包括工作区、接待区、后勤办公室、厨房、餐厅、会议室、休息室和一间"詹姆斯·邦德"室。

　　办公室内着力打造了各种各样的空间，例如合并绿墙、双层挑高空间以及特别的"詹姆斯·邦德"室。

　　"詹姆斯·邦德"室是一间没有窗、像碉堡一样的地方，位于这栋大楼深处，是很难处理的一个空间。而这个看似没有希望的地方已经变成了一间适于安静沉思的贵宾室，打造了一种私密、亲近、平静的氛围。

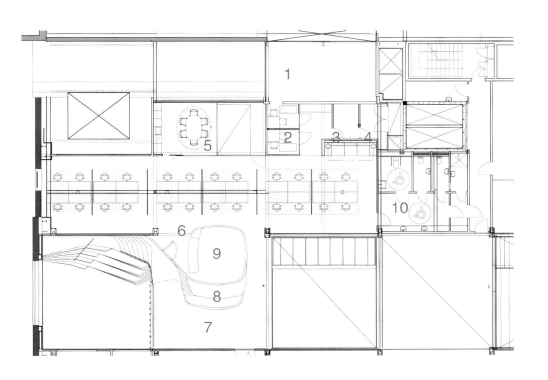

FLOOR PLAN 平面图

1. VENTILATION   1. 通风设备
2. QUIET ROOM/PHONE ROOM   2. 静音室 / 电话室
3. PRINT   3. 打印室
4. DATA   4. 资料室
5. MEETING ROOM   5. 会议室
6. OFFICE   6. 办公室
7. ENTRANCE   7. 入口
8. RECEPTION   8. 接待处
9. BACK OFFICE   9. 后勤办公室
10. WC   10. 卫生间

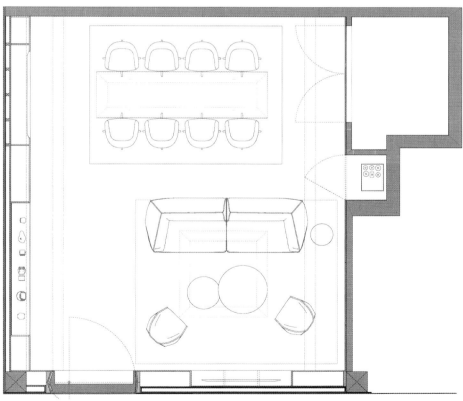

Design Consulting
设计咨询

ARCHITECT 建筑师	Sergey Makhno 谢尔盖·马克诺		
PROJECT AREA 项目面积	141 m²	PHOTOGRAPHER 摄影师	Andrew Bezuglov 安德鲁·贝祖格洛夫

Dizaap Office
Dizaap办公室

▶ CORPORATE CULTURE 企业文化

Dizaap is a young but successful furniture company located in Kiev, Ukraine, they offer a full range of services for construction and repair work, as well as a complete set of houses, apartments, restaurants, offices. Its interior most clearly demonstrates how everyday public space, which we used to call the "office" to turn into a range of ideas and inspiration to work effectively.

　　Dizaap是一家位于乌克兰基辅的年轻且成功的家具公司，他们为建设和修复工作提供全方位的服务，还有一整套住宅、公寓、饭店及办公室的方案。其内部大多清晰地演示了我们每天惯称为"办公室"的公共空间如何变化成一系列可提升工作效率的创意和灵感。

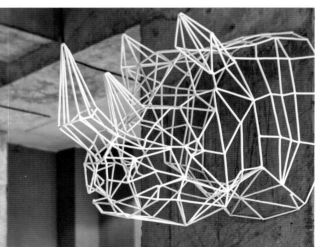

▶ DESIGN CONCEPT 设计理念

By the end of 2014, the company decided to move to a new office. Sergey Makhno found a perfect space. "We had a lot of problems and only one canvas, the size of 141 m² for the hand of the artist. It was necessary to create showroom and office in one space to demonstrate the philosophy, personality and soul of our company".

Inspired interiors studio Sergei Makhno, "Dizaap" decided to avoid hard cosmetics as the leased premises. What have they got? Poured concrete floor, open, but very well-built ceiling wiring, bare walls with highly visible history of creation, eternal, natural, wild tree, aged metal and stone-there are the wholeness of nature and human, which strongly urbanized this world, and only now stopped.

Its project shows how simplicity, minimalism, non-conformism conquer the world. Designers support healthy lifestyle, here they have the Swedish wall, the bike and auger juicer Angel for practicing a raw-food diet. To protect themselves from excessive external noise they have installed sound insulation panels, which consist of the pyramids.

2014年年底，该公司决定迁址到新的办公室。谢尔盖·马克诺发现一个完美的空间。"我们遇到很多问题，并且对于艺术家的手来说，只有一块141m²的画布。在同一个空间打造展示厅和办公室是必要的，这样可以展示我们公司的理念、品性及灵魂。"

Dizaap公司决定让谢尔盖·马克诺这家灵感型室内设计公司设计，以避免把硬装做成租用来的感觉。他们得到了什么呢？灌注混凝土楼面、开放但匀称的天花配线、光秃秃的墙壁，上面有着高可见度的历史，包括创造、永恒、自然、野生树木、老旧金属和石头——这里有人与自然的完整统一，在城市化激烈的世界里，独留一处不同。

该项目展示了简约、极简、不因循守旧是如何征服世界的。设计师遵循健康的生活方式，这里有瑞典风格的墙面、自行车以及螺旋榨汁机Angel，可以培养人们吃天然食物的习惯。为避免过多外界噪音的干扰，他们安装了由角锥体组成的隔音板。

▶ FEATURE 特色

Meeting-room at the same time used as a living room, relaxation area, and sometimes in the evening well-equipped hi-end stereo in addition to the projector it turns into a movie theater.

Bright accents that really "pacified" places are: lemon design chair cocoon of "Ligne Roset" and rocking chairs "Kartel".

The working area is separated from the total area of library shelving, shelves, which flirt sculpture from the Dmitry Greek and Sergey Makhno. A pair of panels of "London Art", a pair of bulky mirrors from "B & B" and "Fendi", a couple of chairs from "Contempo" and a pair of designer tables from "Porada" and "Kristalia" - like that in the interior filled with love and integrity. "A human must be whole, the team must be united."

Everything here is somehow permeated art.

"Do you believe that interior made our lives brighter and wider? I want to work more like working here, create and think positively, even when everything around depressed and fell into a depression." And we believe in healing people's souls through the aesthetics of everyday things, such as office space.

会议室同时作为客厅、休闲区使用，有时在晚上，装备精良的高级音响加上投影仪，它就变成了一个电影院。

真正"使人平静"且点亮空间的地方是"Ligne Roset"柠檬蚕丝椅和"Kartel"摇椅。

工作区域与图书架区是分离的，那些书架摆放着德米特里·格里克和谢尔盖·马克诺设计的雕塑。一副"伦敦艺术"嵌板，一对"B&B"和"Fendi"大镜子，两张"Contempo"椅子，还有两张"Porada"和"Kristalia"的桌子——就像室内充满爱和完整一样。人必须完整，团队必须凝聚。

这里的一切都沉浸在艺术中。

"你相信室内空间会让我们的生活更明亮、更宽广吗？我更想在这里工作，积极地创造和思考，即使有时候周遭的一切都萧条冷清，令人沮丧、意志消沉。"我们相信，通过日常事物的美感可以治愈人们心灵的创伤，比如办公空间。

Others
其他

▲ If you have a dream-creating space
如果您拥有一个创造梦想的空间

▲ You can decorate it as this
可以这样装扮它

▲ Or in this way
也可以这样装扮它

▲ And also in another way
还可以这样装扮它

▲ Here, you can always find your dreamy office space
在这里,您总能找到属于自己梦想的办公空间

Others
其他

Others 其他

PROJECT LOCATION 项目地点	Polanco, Mexico City 墨西哥市波朗科区	DESIGNERS 设计师	Juan Carlos Baumgartner, Gabriel Téllez 胡安·卡洛斯·鲍姆加特纳、加布里埃尔·泰雷兹	DESIGN COMPANY 设计公司	Space Space设计公司	LIGHTING 照明设计	LUA LUA公司
PROJECT AREA 项目面积	4800 m²	PHOTOGRAPHER 摄影师	Paul Czitrom 保罗·奇特罗姆	FURNISHING 家具设计	Herman Miller 赫曼·米勒公司	DESIGN TEAM 设计团队	Marcos Aguilar, Humberto Soto, Diana Casarrubias, Yuri Rodríguez 马科斯·阿吉拉尔、汉姆托·索托、黛安娜·卡萨如贝亚斯、尤里·洛迪古斯

CP Group
CP集团办公室

▶ CORPORATE CULTURE 企业文化

CP Group is a leader company in insurance with more than 35 years of experience in management consulting and comprehensive risk in the Mexican market and abroad. The company has all type of insurance like car insurance, bus insurance, damage and bond.

　　CP集团是保险行业的主导企业之一，在墨西哥及国外市场拥有超过35年的管理咨询和全面风险管理经验。该公司涵盖各类保险业务，例如汽车保险、公共汽车保险、损害保险和债券保险。

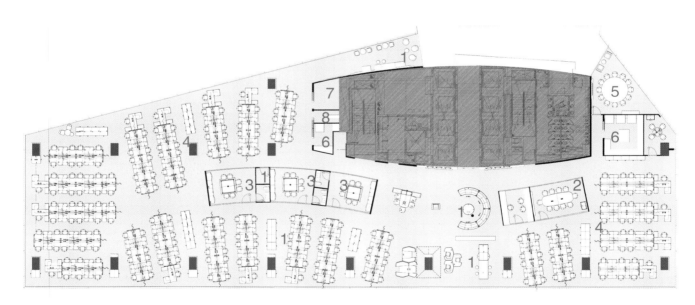

15TH FLOOR PLAN 15层平面图

1. CASUAL AREA
2. MEETING ROOM
3. PRIVATE DIRECTOR
4. OPEN AREA
5. TECHNOLOGY ROOM
6. COPY/PRINT
7. ELECTRIC ROOM
8. IDF

1. 休闲区
2. 会议室
3. 私人助理办公室
4. 开放区域
5. 技术室
6. 复印/打印室
7. 电气室
8. 综合资料文件室

▶ DESIGN CONCEPT 设计理念

The corporate offices of the CP Group are located on the fifteenth floor of the Terret Building, in Mexico City. Pentagono Estudio participated with specialized furniture, graphic arts and applications, thus generating an architectural balance in conjunction with the identification of spaces and visual communication. Each level incorporates particular details that generate an integral thematic concept related to the company's business areas: City, Health and Automotive.

  CP集团企业办公室位于墨西哥城特雷特大楼的第十五层。Pentagono Estudio公司参与了专业家具、平面艺术与应用程序的设计，从而生成一个空间身份识别与视觉传达结合的建筑平衡效果。每个层级都包含了具体的细节，创造出一个与公司商业领域相关的完整主题概念：城市、健康和汽车。

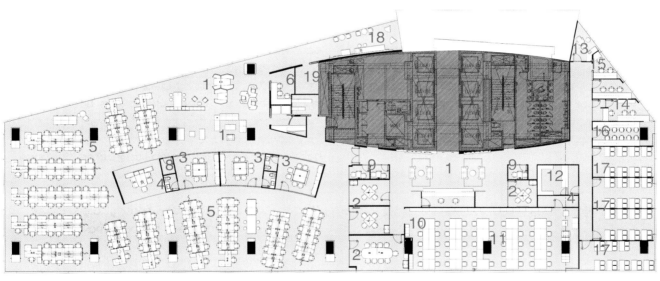

16TH FLOOR PLAN 16 层平面图

1. RECEPTION	8. STORAGE	14. CASH AREA	1. 接待室	8. 储藏室	14. 现金区
2. MEETING ROOM	9. ONE TO ONE	15. INDEMNITY	2. 会议室	9. 一对一室	15. 赔偿区
3. PRIVATE DIRECTOR	10. LOCKERS	16. AUDITORS	3. 私人助理办公室	10. 储物间	16. 审计区
4. MEDIA SPAC	11. POOL	17. TRAINING	4. 媒体区	11. 水池	17. 培训区
5. OPEN AREA	12. MAIL BOX	18. CASUAL AREA	5. 开放区	12. 信箱区	18. 休闲区
6. HELP DESK	13. NURSING	19. ELECTRIC ROOM	6. 咨询台	13. 护理室	19. 电气室
7. COPY/PRINT			7. 复印/打印室		

▶ FEATURE 特色

The corporate interior project of the CP Group consists of three levels; the architectonic program includes spaces such as vestibule, reception, open areas, private areas, meeting rooms, executive dining rooms, informal or one-to-one meeting rooms or training, areas of support, IDFs, technology room, digitalization room, cashier's office, among others. Floor 15 is private, the greater part being open. Floor 17 is also private, with the holding area (management area), and floor 16 is public with a training area.

The client requires the offices to obtain LEED certification, and the design should therefore take into account the energy efficiency and sustainability parameters.

Wall video is located on the core walls accessing the elevator, for the purpose of projecting company marketing to users, clients and visitors.

The reception area is on floor 16, and has recovery board facing on the solid surface furniture. In the background is a metallic plate-lined wall showing the logotype of the CP Group, for providing service at the three levels. There is a waiting area and a services area for outside visitors; control is by glass doors with scanners permitting user access.

The private directors' rooms consist of an open area with audiovisual equipment, screens on the walls and, in some, additional spaces with projectors and electrical screens.

On floor 17, in the holding area, there are three executive dining rooms, lounge, secretarial area, directors' offices, board room, and a kitchen, with dressing room and bathroom facilities, open areas and supporting areas.

Located on floor 16 is the reception area, waiting space, training rooms, cashier's office, nursing area, compensation, treasury, meeting and multiple usage rooms, open area, supporting areas and cashier, among others.

On floor 15, the technology, printing, digitalization room, open area, supporting areas, etc., are to be found, having screens on the columns in open areas.

　　CP集团内部项目包括三个等级：建筑项目包括前厅、接待室、开放区域、私人领域、会议室、高管餐厅、非正式或一对一会议室，还有培训室、支持区、IDF区、技术室、数字化房间、出纳办公室等。15层是私人领域，大部分是开放的。17层也是私人领域，设立等候区（管理区域），16层有一个培训区，完全公开。

　　客户要求这些办公室取得绿色建筑认证，所以设计应考虑能源效率及可持续性参数。

　　视频墙位于电梯附近的主墙壁上，为了向用户、客户及访客进行公司行销。

　　接待区在16层，朝固体表层家具的是复原板。金属板墙作为背景，展示了CP公司的商标，以三个层次提供服务。对外部访客开放的等候区、服务区；通过带有扫描器的玻璃门控制，允许访客进入。

　　私人董事室由一个开放区域构成，包括试听设备、屏幕墙、一些附加空间，还有投影仪和电子屏幕。

　　17层是等候区，有三间高管餐厅、休息室、秘书室、董事办公室、董事会议室、厨房、更衣室、浴室设施、开放区域以及支持区。

　　16层是接待区、等候区、培训室、出纳办公室、护理室、薪酬室、财政部、会议室、多功能室、开放区域、支持区和出纳室等。

　　15层可以看到技术室、印刷室、数字化房间、开放区域、支持区等，开放区域的廊柱上都有屏幕。

There are facilities located between the informal meeting rooms and the kitchen which permit garbage to be separated by organic and inorganic.

All floors have a closed printing room with extractor, together with supporting furniture for stationery storage. The one-to-ones are small cabins with a counter, a chair and telephone line, allowing two users to communicate with much more privacy. Meeting rooms will be equipped according to need and size. Viewing will be by projector or screening.

Features which make these spaces stand out are the strategically-located meeting rooms. These offer coffee and water services, together with an informal space at which presentations can be made, utilizing either a laptop or media scope.

Every informal meeting space is different and has a different subject matter, designed to reinforce the intention to include something special on each floor.

Soffits in the open area are circular lintels suspended from the slab; the private offices have plaster board soffits with white glass elements. Private meeting rooms possess presence detectors and diffusers. In the open area, circular lights suspended at random were proposed by the lighting designer.

非正式会议室和厨房之间有一些设施，可以让垃圾进行有机和无机分离。

每层楼都有一个封闭的印刷室，里面有提取器以及存储文具纸张的支持装置。一对一的小包厢有一张柜台、一把椅子和电话线，允许两人进行更加私密的交流。会议室按照需求和大小进行配备，通过投影仪或屏幕来观看。

让这些空间突出并引人注目的是处于战略性位置的会议室。这些会议室提供咖啡和供水服务，还有一个非正式空间可以用笔记本电脑或媒体来做业务陈述。

每一个非正式会议空间都是不同的，有不同的主题，旨在加强每层楼包含一些特别之处的意向。

开放区域的拱腹采用圆形门楣，从平板上悬吊下来；私人办公室拥有灰泥板拱腹和白色玻璃元素。私人会议室拥有车辆占用轨道检测器以及柔光镜。在开发区域，圆形灯的灯光随意摆动着，这是灯光设计师的提议。

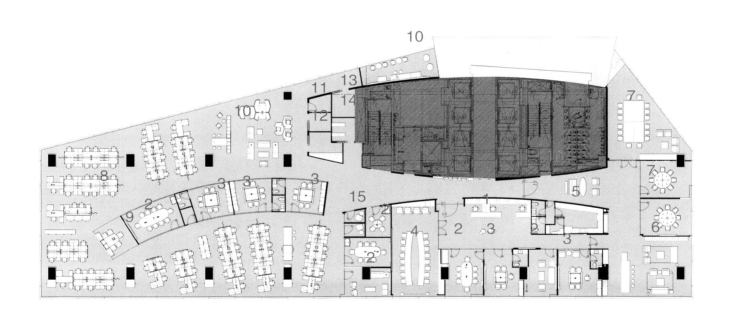

17TH FLOOR PLAN 17层平面图

1. RECEPCTION	6. LOUNGE	11. STORAGE	1. 接待室	6. 休息室	11. 储藏室
2. MEETING ROOM	7. DINING	12. COPY/PRINT	2. 会议室	7. 餐厅	12. 复印/打印室
3. PRIVATE DIRECTOR	8. OPEN AREA	13. ELECTRIC ROOM	3. 私人助理办公室	8. 开放区域	13. 电气室
4. COUNCIL MEETING	9. MEDIA SPACE	14. IDF	4. 理事会议室	9. 媒体区	14. 综合资料文件室
5. KITCHEN	10. CASUAL AREA	15. PHONE BOOTH	5. 厨房	10. 休闲区	15. 公用电话间

Others 其他

PROJECT LOCATION 项目地点	Istanbul, Turkey 土耳其伊斯坦布尔	DESIGNERS 设计师	Serhan Bayik, Okan Bayik, Ozan Bayik, Oya Canatar, Erhan Arslan, Tugba Kılınç 塞尔汗·巴依克、奥坎·巴依克、奥赞·巴依克、奥亚·卡那塔、埃尔汗·阿尔斯兰、图巴·基林		
PROJECT AREA 项目面积	2250 m²	PHOTOGRAPHER 摄影师	Gürkan Akay 居尔坎·阿凯	DESIGN COMPANY 设计公司	OSO Architecture OSO建筑公司

Met Global
Met全球公司

▶ CORPORATE CULTURE 企业文化

Met Global office project is located within an area of 2.250m² in Davutpasa Campus of YTU. The office located at a single space on the top floor has a capacity of 325 personnel. The business field where the company operates manufactures software and information technologies on tourism basis. In this sense, they have a quite young and dynamic staff integrated with the digital environment. Created moving from this point, the interior concept setup is designed to be matched with the general characteristic of the office and the personnel profile.

Accordingly, the modern and contemporary design language as adopted is blended with the color use, and vitality and dynamism are brought to the interior spaces. Moreover, there is a reception wall that shapes the first impressions of the users and visitors about the brand at the entrance hall of the office. In the design of this wall, there is inspired from the 'pixel" form with respect to the software and information industry as produced by the company.

Met全球公司办公室项目坐落于伊尔德兹技术大学达瓦帕萨校区，占地面积2250m²。该办公室位于顶层的单一空间，可容纳325人。该公司涉及的商业领域基于旅游业、生产软件和信息技术。从这层意义上说，他们拥有可以与数字化环境相结合的年轻、活力充沛的员工。然而创造性地脱离这一点，室内理念被设计为符合办公室整体特点及人员简历。

于是，采用现当代的设计语汇与色彩运用相融合，将生机和活力带入室内空间。此外，一面接待墙给使用者和访客留下了关于办公室入口大厅标牌的第一印象。在这面墙的设计中，带着对该公司生产的软件和信息工程的崇敬，激发了"像素"形式的灵感。

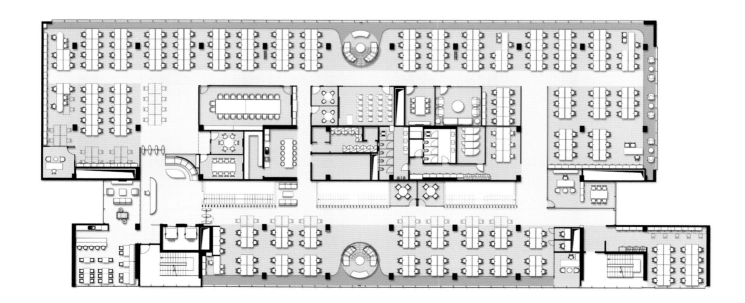

▶ DESIGN CONCEPT 设计理念

OSO Architecture is an Istanbul-based architectural studio established by Okan Bayık, Serhan Bayık and Ozan Bayık in 2007. A strong emphasis is given to the critical design process within the studio; they resist predetermining architectural solutions to a client's brief prior to a thorough investigation of each project's unique situation.

Their criteria for design is to pay close attention to contemporary design methods, new materials and the economic considerations of the client. They believe that a good project must combine all these areas. Moreover, the most important thing for them is to look at a project from a new perspective.

In order to achieve these aims, OSO architecture has brought together three different professions; architecture, interior design and civil engineering. OSO Architecture specialises in architecture, interior design and project management. Their aim is to create examples which are not only unique, but also combining efficiency, economy and design.

  OSO建筑公司是一家伊斯坦布尔建筑工作室，2007年由奥坎·巴依克、塞尔汗·巴依克和奥赞·巴依克创办。该工作室重点强调严格的设计过程；他们拒绝根据客户的简要描述预先确定建筑方案，而是要在这之前对每个项目的独特情况进行彻底调查。

  他们设计的标准是密切关注当代的设计方法、新型材料及客户的经济因素。他们认为一个好的项目必须综合所有这些领域。此外，对他们来讲最重要的便是以一种新视角来审视一个项目。

  为实现这些目标，OSO建筑公司将建筑、室内设计和土木工程三个行业交融在一起，其专门从事建筑、室内设计和项目管理。他们的目标是打造一些不仅特别而且结合了效率、经济及设计因素的案例。

▶ FEATURE 特色

At the office where the general layout setup is planned as an open office, the recreation areas are integrated into the open office. Thus, it is encouraged to increase the internal communication at the office and to trigger the creative ideas. To this end, grass platforms are created at the edges of the exterior of the open office, and these spaces are connected with the recreation space (gazebo) in the mid-venue. Furthermore, there is an independent café in the office, and a playroom connected with this venue is also planned.

On the other hand, to eliminate some disadvantages caused by the open office setup, and especially to protect the privacy as required occasionally, there are planned 2 "think rooms". It is aimed to enhance the concentration of the users by creating a completely enclosed and isolated environment in these spaces. In addition, in planning setup, the number of the meeting rooms is maximized, and some rooms are transformed into the "informal" meeting rooms in "lounge" style.

Thus, multifunctional spaces are created, where both meetings are held, and "brainstorming" can be made. In general sense, matching and integrating this office space designed with their own company identity, field of business, and user profile are determined as a basic design problem, and thus, it is aimed to bring sense of belonging to the space.

该办公室总体布局设置为开放空间，加入了娱乐区域。因此，鼓励增加办公室内部联络，激发创意灵感。为了达到这个目的，在开放式办公室的外部边沿打造了草坪平台，这些空间连接了中部位置的娱乐区域。此外，该办公室内设立了一个独立的咖啡室，还有一个连接此中部位置的游戏室正在计划。

另一方面，去除开放式办公设置引起的一些劣势，尤其是偶尔根据需要保护隐私，打造了两间"思考室"，旨在通过在这些空间打造一个完全封闭、隔离的环境来提升使用者的专注力。另外，在平面图设置中，扩大了会议室的数量，一些房间转变成"躺椅"型的"非正式"会议室。

因此，打造了可以举办会议也可以进行"头脑风暴"的多功能空间。一般来讲，匹配并整合这个按照该公司特性、商业领域及用户信息而设计的办公空间是一个非常根本的设计难题，因此，为该空间带入归属感就成为了目标。

Others 其他

DESIGNER 设计师	Jean de Lessard 珍·德·莱萨德	PHOTOGRAPHER 摄影师	Adrien Williams 阿德里安·威廉姆斯
PROJECT AREA 项目面积	418 m²		

BICOM Communications
BICOM公司办公室

▶ CORPORATE CULTURE 企业文化

Guided by a young and modern vision, BICOM Communications is an established public relations firm specializing in subjects relating to lifestyle. Based in Montréal with offices in Toronto and operations throughout Canada, BICOM offers a comprehensive range of customized press relations, event planning, and marketing-communication services. Its mission? To effectively fulfill client objectives in terms of visibility and strategic positioning. Its team has a twenty five talented professionals whose common objective is not just to reach, but to exceed all expectations thanks to an incredibly proactive approach.

BICOM通信是一间已正式成立的公共关系公司，以年轻时尚的视角作为指引，擅长与生活方式相关的主题。BICOM通信公司总部位于蒙特利尔，办公室均位于多伦多，经营范围覆盖整个加拿大，提供综合性定制印刷业务、活动策划以及市场营销通信服务。其使命是依照可见性和战略定位有效地满足客户需求。其团队拥有25名专业人才，运用一种非常积极主动的方法，他们的共同目标不仅仅是达到，而是超越所有期望值。

FLOOR PLAN 平面图

1. TOILETTES　　1. 卫生间
2. CUISINE　　　2. 餐厅

▶ DESIGN CONCEPT 设计理念

Designing a space as a creative medium and productive community is the intention Jean de Lessard, a creative designer has manifested through form for the BICOM offices design.

"They are young people of the new generation, smart people. My design strategy is reflecting their agency and the method of operation. The diverse personalities of the 24 employees are also shining through, " explains interior designer Jean de Lessard about BICOM.

The design, in the words of the designer, more of a controlled chaos, makes use of open space in this vast expanse which was created by the previous joining of contiguous office spaces to meet the agency's continuous expansion. The integration of several private offices to accommodate eventually 35 people plus a few open areas dedicated to social interactions between colleagues has thus become a reality, thanks to chaos. This is so because the fact of being together should be equally inspiring than inspired, why not then communicate between colleagues in a fun, functional and sophisticated environment.

珍·德·莱萨德的目的是设计一个可作为创意媒介和生产共同体的空间，她是一名创意设计师，通过BICOM公司的办公室设计得以显现。

关于BICOM这个案子，室内设计师珍·德·莱萨德说道，"他们是新一代的年轻人，是聪明的人。我的设计策略是反映他们的机构及经营方法。24名员工不同的个性也在其中闪烁着光亮。"

用设计师的话来说，该设计更像是一团受约束的混沌之物，早前，为满足公司的持续扩张，在邻近办公空间打造出这个广阔区域，使该区域的开放空间得以利用。由于混沌，一些可容纳35人的私人办公室与促进同事间社交互动的几个开放区域的融合最终成为了现实。确实如此，人们在一起的时候，激励与被激励应该同等重要，那么为什么不让同事们在一个有趣、功能性强、精致的环境下交流呢。

▶ FEATURE 特色

This is the case with the BICOM project: a systematic deconstruction of a place and functions before its reconstruction as a stylized village, surely, but in a totally offbeat way.

Jean de Lessard chose the archetype of the small house, because it brings out emotions that trigger, for many, happy memories at the summer cottage by the lake and campfires at night. The houses are positioned around a few public areas. Despite the rigidity of the construction materials used to build them the cottages are a flexible design system that allows growth within the agency.

Volumes are simplified, lines are pure and masses of colour are used parsimoniously to form a warm and cozy architectural envelope against the stark universe of the bare white walls and high ceilings of the old factory. Further, the seemingly erratical positioning which also defines circulation axis is aiming at breaking monotony and to encourage interactions between people.

　　这是BICOM公司的项目：在重建为乡村风格之前，以一种完全另类的方式，进行一个地方及其功能的系统解构。

　　珍·德·莱萨德选择了小型屋宇的原型，因为它带来了一些情绪情感，引发湖边夏季小别墅及夜晚篝火旁很多快乐的回忆。这些房子围绕在一些公共区域四周。不管用于建造它们的施工材料质地多坚硬，这些小别墅形成了灵活的设计系统，允许公司内部发展。

　　量体被简化，线条被纯化，大量色彩得以运用，靠着老工厂荒凉的地域，光秃秃的白墙和高高的天花，形成一个温暖舒适的建筑包膜。此外，看似游离的配置定义了循环轴线，目的是打破千篇一律，并鼓励人们之间的互动。

Others 其他

PROJECT LOCATION 项目地点	Shanghai, China 上海	DESIGNER 设计师	Andy Leung 梁裕能	DESIGN COMPANY 设计公司	ISSI DESIGN LTD 一思室内设计咨询事务所
PROJECT AREA 项目面积	2216 m²				

Blinq Office
缤客办公室

▶ CORPORATE CULTURE 企业文化

Blinq Trading Co., Ltd was founded in 2011, which is a leading O2O whole marketing company throughout the country, devoted to help clients solve whole marketing programs via on-line and offline channels. It mainly engaged in the cooperation with Shanghai Shentong Metro Group and Shanghai Metro transportation system to develop a new retail platform based on transportation network.

缤客商贸（上海）有限公司成立于2011年，Blinq是一家全国领先的O2O整体营销公司，通过线上线下渠道，帮助客户解决整体营销方案问题。主要是与上海申通地铁集团及上海地铁交通系统，共同开发基于交通网络的全新零售平台。

FLOOR PLAN 平面图

1. TRAIN MODEL　　1. 火车模型
2. UPSLOPE　　　　2. 上坡

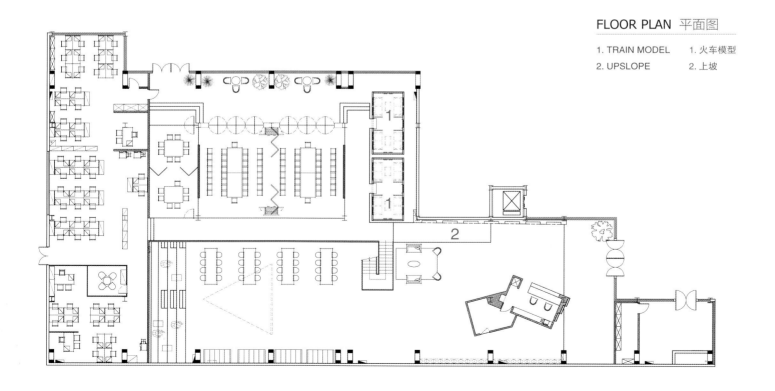

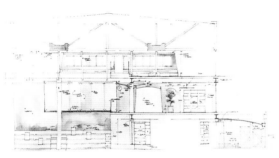
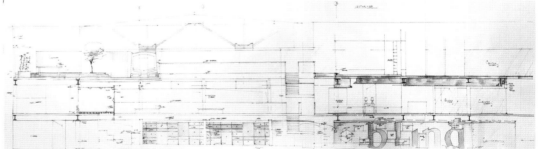

▶ DESIGN CONCEPT 设计理念

As Blinq is an international O2O metro brand, ISSI Design hopes to express a corporation image of wisdom, vitality and happiness in its new office of Shanghai. In the old factory building with three floors and a total area of 3000 square meters, the designer uses "journey" as the concept to show a workspace full of humanness and a new working mode through different stations and the connection among the scenes both inside and outside the room. Various scenes combine together, such as the basketball court, the hanging garden, library and so on, creating an energetic and open working environment via the reorganization of colors and raw steel structures.

Blinq作为一个国际化的地铁O2O品牌，ISSI希望在其新上海办事处表现出智慧、活力及欢乐的企业形象。在楼高三层总共3000m²的旧厂房，设计师以"旅程"作为概念，通过不同的站台，透过车窗内外的场景串联，展示出一个充满人性化的工作空间及一种新的工作模式。利用不同的场景配合，例如篮球场、空中花园、图书馆等，透过色彩和原钢结构的重组，创造出一个活力而且开放的工作环境。

FLOOR PLAN 平面图

1. TRAIN MODEL   1. 火车模型

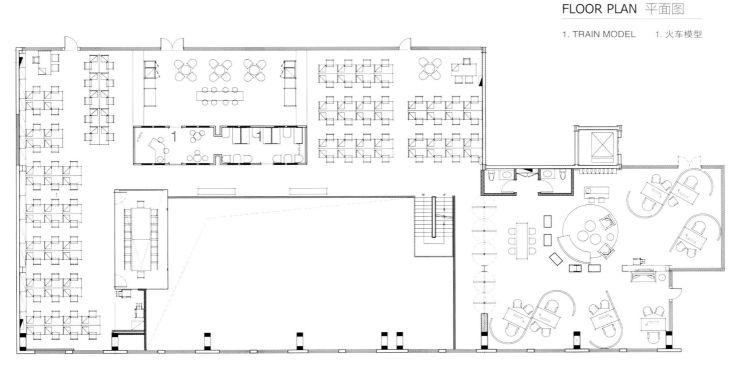

FLOOR PLAN 平面图

1. CONTAINER 1. 集装箱

▶ FEATURE 特色

On the plane layout, youth, originality and cutting-edge fashion are defined; open workplaces are planned to make it closer for the communication among companions; the high-ceilinged area in the atrium is centered on the third floor of the office, while the reception area and meeting rooms on the first floor as well as the working areas on the second and third floor are defined to be dynamic and static; each department is adjoined with others on the function, which makes the team work connected seamlessly in the space; while in the interior, paint in different colors, high-class carpets, riffled plate, tempered glass and other materials echo with each other, creating the space into a fashionable area.

在平面布局上，以年轻、创意、前沿时尚为定义；多以敞开式办公区而规划，使小伙伴们沟通零距离；三层办公空间以中庭挑空区为中心，一层的接待、会议与二、三层办公区定义为动与静；在功能上每个部门相互为邻，在空间上将团队合作无缝连接；室内采用了不同颜色油漆、高级地毯、花纹钢板及钢化玻璃等材质的互相映衬，将空间打造成时尚之区。

Others 其他

PROJECT LOCATION 项目地点	Stockholm, Sweden 瑞典斯德哥尔摩	DESIGNERS 设计师	Lisa Bruch, Ola Hermansson 丽莎·布鲁赫、奥拉·赫尔曼松	DESIGN COMPANY 设计公司	MER MER设计公司
PROJECT AREA 项目面积	2700 m²	PHOTOGRAPHER 摄影师	Måns Berg 芒·贝格		

Nexus
Nexus办公室

▶ CORPORATE CULTURE 企业文化

Nexus is one of the world leaders in security solutions. They moved from three different locations to Telefonplan, Stockholm. Nexus is undergoing an expansive phase and needs bigger premises while the move is meant to weld together the various divisions within the Nexus. Also, Nexus move from cellular offices, larger rooms to an open plan divided into different security classes.

Nexus是世界级安全解决方案的领导者之一。他们从三个地方迁入斯德哥尔摩的创意园区。Nexus正在经历一轮扩张阶段，需要更大的办公空间，同时，本次搬迁有意要把Nexus众多部门联结在一起。并且，Nexus从分格式办公室转变为更大房间以及可分为不同安全级别的开放式格局。

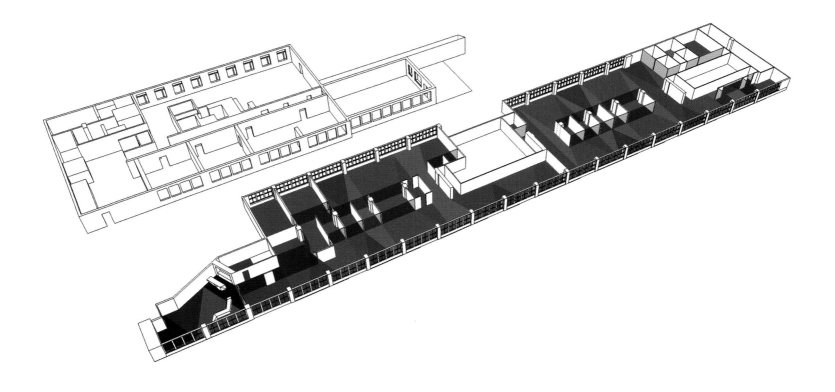

▶ DESIGN CONCEPT 设计理念

The concept is based on the Nexus graphic manual and logo. The X represents the junction between the different activities within the company and the expertise and experience. The result is an office that feels modern, Scandinavian while still respecting the original architecture.
Nexus operation is partly SCRUM based. This means that they need flexible solutions for workstations when teams change in size and composition. To support that, designers have developed the idea of a mobile meeting place.

设计理念基于Nexus的图标指南及商标。X代表公司内不同活动与专业知识经验之间的联结。从而形成了一个令人感到现代、北欧风同时尊重原始建筑的办公室。Nexus的运作在一定程度上基于SCRUM。这就意味着当团队改变规模或构成的时候，他们在工位方面需要灵活的解决方案。为了支持这一点，设计师提出了移动型会议地点的构想。

▶ FEATURE 特色

The telephone factory is Ericsson's old office at Telefonplan, a building with historical interest from the 30`s, that now houses Nexus in three floors. The first floor houses the production area and the 2nd floor is the main office floor with about 100 workstations. The top floor, Ericsson´s old reception rooms with original features, serves as a meeting place with lunchrooms, large meeting rooms, patio, relax area and showroom for product demonstrations.

这间电话工厂是爱立信公司在创意园区的办公旧址，这栋楼是30年代的历史遗迹，现今成为了Nexus公司的三层办公楼。一层是生产区，二层是拥有将近100个工位的主要办公区。顶层是爱立信公司从前的接待室，带着些许原始特色，现在作为会客区，其中包括午餐室、大型会议室、天井、休息区以及产品展示间。

PROJECT LOCATION 项目地点	Moscow, Russia 俄罗斯莫斯科	ARCHITECTS 建筑师	Pedra Silva Architects, Luis Pedra Silva, Maria Rita Pais 佩德拉·席尔瓦建筑事务所、路易斯·佩德拉·席尔瓦、玛丽亚·丽塔·派斯		CARPENTRY AND MASONRY 木工、石工	BEC		
PROJECT AREA 项目面积	3370 m²	PHOTOGRAPHER 摄影师	Fernando Guerra 费尔南多·格拉	GRAPHIC DESIGN 平面设计	P06	CONTRACTOR 承包商	VURAL	PARTITIONS 分区	OWD
DESIGN TEAM 设计团队	André Góis, Dina Castro, Hugo Ramos, Hugo Ferreira, João Alves, Paulo André, Ricardo Sousa 安德烈·戈伊斯、迪娜·卡斯特罗、雨果·拉莫斯、雨果·费雷拉、若昂·阿尔维斯、保罗·安德烈、里卡多·苏萨								

Uralchem Headquarters
Uralchem总部

▶ CORPORATE CULTURE 企业文化

Uralchem Group is the largest ammonium nitrate producer in Russia and the second largest ammonia and nitrogen fertilizer producer in Russia. Uralchem is the company that respects and values its employees, shares their energy, experience, and expertise with it, cares about their health and safety at work and offers good employment benefits and opportunities for career development. Wherever its employees' responsibilities lie, they are united by one common goal which makes the Company's work fruitful and interesting.

Uralchem集团是俄罗斯最大的硝酸铵生产商和第二大氨氮肥料生产商。Uralchem尊重和珍视公司的员工，乐于分享能量、经验和专业知识，关心他们在工作中的健康和安全，同时提供良好的就业福利和职业发展机会。无论员工的责任是什么，他们通过一个共同的目标团结在一起，这个目标就是让公司的工作富有成效且具有趣味性。

▶ DESIGN CONCEPT 设计理念

The brief was to create a space that would enable ideal working conditions for staff while also reflecting the company's dynamic, relaxed and youthful spirit. The project space occupies an entire floor of Imperia Tower, a skyscraper in Moscow City.

The first challenge arose when initially visiting the site and realizing that some of the tower floors were previously built to be a hotel. This explained the short floor to ceiling height and inexistence of raised floors that would be ideal for office use. So the design team set out to answer the question…"How do they make a space look taller than what is actually is?" So how do they make a space look taller?

While searching for an answer to this question, designers ended up solving their initial problem and a series of others by rethinking the ceiling. They created an innovative and unique ceiling that not only makes the space feel larger but also removes unappealing elements that make up the traditional ceiling.

The result is a horizontal "visual filter" that is laid out on a matrix made up of discs where lighting and air outlets are placed according to need. Behind this filter designers have all the technical mechanisms that are hidden from sight such as fire detectors, air ducts and other gizmos.

核心理念是创造这样一个空间，既能保证员工理想的工作环境，同时也反映出公司的活力、轻松及朝气。该项目空间占据因佩里亚大厦的一整层，该大厦是莫斯科市的一座摩天大楼。

首次访问该地区，发现此大厦中有些楼层原先是用于建造酒店的，那么第一个挑战就出现了。这也解释了地板到天花板的高度，以及那些活地板，非常适合办公使用。于是，设计团队开始着手回答这个问题……"如何让空间看起来比实际上更高？"

那么他们怎样让空间看起来更高呢？

在不断地寻找该问题答案的过程中，设计师们通过重新考虑天花板，解决了原先的问题，还解决了其他一系列问题。他们设计出一个新颖独特的吊顶，不仅在空间上感觉较大，也消除了一些为装扮旧天花板而不具任何美感的元素。

其结果是一个水平"视觉滤波器"，即布置在由圆盘所组成的矩阵上，其中的照明和送风口可根据需要进行放置。在该过滤器的背后，设计师们设置了科技机制，且都是隐藏在视线之外的，例如火灾探测器、空气管道和其他小玩意等。

▶ FEATURE 特色

The ceiling is made up of white circular elements that form a continuous surface embracing the space and reflecting natural light. Artificial light is achieved by clicking in light discs according to the amount of light required around a particular space. You do the same by placing equivalent diameter air-conditioning vents that become fully integrated in the ceiling system. The result is a flexible system of interchangeable suspended disks, allowing for easy access to the upper infrastructure while minimizing the effect of a lower ceiling.

If designers had used a regular office ceiling, the space would have felt cramped and claustrophobic, but instead it feels big and airy while still ticking all the boxes of the performance you need from a regular office ceiling. From initial prototypes in Innsbruck, Austria to actual production in Ankara, Turkey the result is a bespoke answer to the initial problem resulting in the main aesthetic element of this space.

As for the space itself, the office is arranged around the central service core of the building, working as a distribution nucleus between departments. The circulation route around this core is emphasized by a continuous wood surface that is randomly cut so it secretly hides storage units and doors.

Besides the aesthetic nature of the space, key spaces were provided for staff to promote well-being and productivity. Spaces such as silent rooms to improve concentration, small rooms for unplanned meetings, noise removing elements for the open space and a large coffee lounge where staff mingle and share their experiences.

The space is occasionally interrupted by "glass bubbles" that contrast in nature to surrounding circular references. These "bubbles" contain team leaders and noisy rooms within a sound proof environment. The pictograms add a splash of colour in an otherwise calm and neutral environment.

A project with bespoke solutions, from ceiling to special spaces culminating in a comfortable working environment that promotes happiness and productivity.

天花板由白色圆形元素组成，形成了一个连续的平面，既揉合了空间，又折射出自然光。根据特定空间所需光量，敲击光盘以获得人造光。放置当量直径的空调通风口，在天花板系统中实现完全集成，同样可实现上述需求。其结果是一个可互换悬浮磁盘系统，可方便使用较高的基础设施，同时降低位置较低的天花板的影响。

如果使用普通的办公室天花板，空间上会令人感到局促和幽闭。但反过来，感觉到宽敞通风，虽然滴答作响，也满足了人们对普通办公室天花板的所用性能的需求。从最初在奥地利因斯布鲁克的原型到在土耳其安卡拉生产，其结果就是对最初在此空间内美学因素导致的问题的最佳答案。

至于空间本身，办公室围绕着中央服务核心进行布置，作为部门之间的分配核心。围绕着该核心的循环路径由一个连续的木质平面进行突出，循环路径可随机切断以便秘密地隐藏起一些装置和大门。

除了空间的审美本质，该项目也可向员工提供有益空间，以促进身体健康，提高生产力。不同空间的使用，作用大不同。例如，无声室有助于集中注意力，小房间则可用于突如其来的会议，可将消除噪声的因素用于开放的空间里，而一个宽敞的咖啡厅可使员工们打成一片，分享生活。

空间偶尔会被"玻璃气泡"干扰，在性质上与周围的循环参照物相对照。这些"气泡"包含团队领导以及在一个隔音环境下的嘈杂的房间。

具备专门解决方案的项目，从天花板到一些特殊空间都无微不至，最终以实现舒适的工作环境，从而提升了员工的幸福感，提高了生产力。

Others
其他

PROJECT LOCATION 项目地点	Moscow, Russia 俄罗斯莫斯科	CHIEF ARCHITECT 首席设计师	Boris Voskoboynikov 鲍里斯·沃斯科博伊尼科夫	DESIGN COMPANIES 设计公司	VOX Architects, Boris Voskoboinikov Architectural Studio VOX建筑公司、鲍里斯·沃斯科博伊尼科夫建筑工作室
PROJECT AREA 项目面积	1200 m²	PHOTOGRAPHER 摄影师	Alexey Knyazev 阿列克谢·克尼亚杰夫	LIGHTING DESIGNER 灯光设计师	George Kelekhsaev 乔治·克勒克赛弗
PROJECT TEAM 项目团队	Olga Ivleva, Elena Merzalova; Sergey Kurepin (Project Engineer), Eugeniy Kiselev (Project Manager) 奥尔加·艾维莱瓦、埃琳娜·梅尔察洛瓦、谢尔盖·库雷宾（项目工程师）、欧金尼·基谢廖夫（项目经理）				

E:MG Office
E:MG办公室

▶ CORPORATE CULTURE 企业文化

E:MG is a BTL agency in Russia, their main product is effectiveness, which they understand as proven ability to deliver such expertise and services, that empower their client to sell more smarter: do it in a more affordable, faster and consumer-friendly way. Its clients include Kraft Foods, Mercedes-Benz, Coca Cola, Mirosoft, Philips etc.

　　E:MG是俄罗斯一家线下代理公司，他们的主要产品是时效性，在提供此类专业技术与服务方面，他们认为自己的这种能力已经得到证明，这使他们的客户更精明地进行销售：用一种更实惠、更迅捷、更方便消费者的方式。他们的客户包括美国卡夫食品、奔驰汽车、可口可乐、微软、飞利浦等。

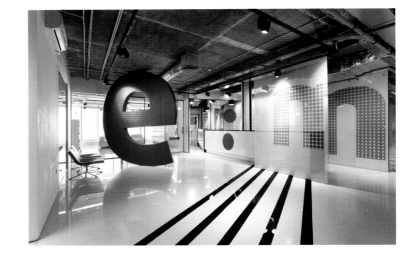

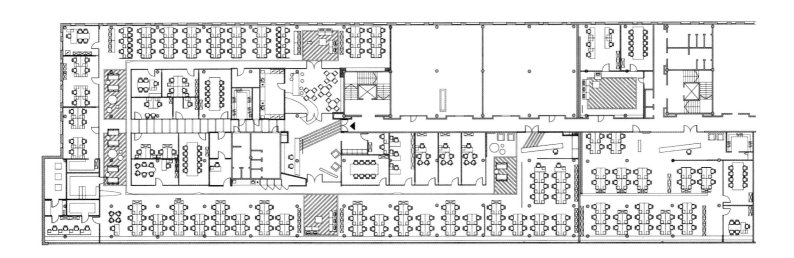

▶ DESIGN CONCEPT 设计理念

When designing an office for E:MG BTL agency, Boris Voskoboynikov put his choice in favor of common materials in the interior.
"Imagery of the interiors is built around graphic elements. Most significant is a large spatial composition that comprises E, M, G letters created of different materials ."—comments by Boris Voskoboynikov.

  为E:MG线下代理公司设计办公室的时候，鲍里斯·沃斯科博伊尼科夫倾向于选择常用的材料来进行室内设计。
  "室内意象围绕着图形元素建立。最显著的是包含不同材质的字母E、M、G的大型空间布局。"——鲍里斯·沃斯科博伊尼科夫

▶ FEATURE 特色

Its most valuable feature is that it 100% corresponds to the activities performed by the company, besides, it boosts the creativity of the company employees. Simple materials and techniques were implemented in the decoration: self-leveling flooring, central volumes – the distinctive feature of the author, so-called "architecture in architecture" – are made of plaster slabs and paint-coated. At the same time, the client was receptive to the recommendations of Boris Voskoboynikov and the light designer: the lightning system installed in the office turned to be the most appropriate. The largest part of the office a public open space, although some zones, such as for an accounting department, are isolated.

When locating the decorative text, the architect was playing with the perspective: he placed the letters on different surfaces and at a certain point one can read the company's name.

Quotation: "In every office project we try to organize the working places location—that tends to be chaotic most of times in correct structures: this really changes the attitude people have towards their own job."

其最有价值的特色是它100%符合该公司执行的行为活动，此外，它大大提高了员工的创造力。装修方面应用了简单的材料和技术：自流平地板、中央量体——设计师的独特风格，即所谓的"建筑中的建筑"——由石膏板和油漆颜料包饰。同时，客户接受了鲍里斯·沃斯科博伊尼科夫及灯光设计师的建议：为办公室设置了最适合的灯光系统。该办公室最大的部分是一个公共的开放空间，尽管有一些区域是隔离的，比如财务部。

当设置装饰文字时，建筑师运用了透视图案，他把字母放在不同的表面，于是在特定的角落人们可以看到公司的名字。

引文："每个办公室项目，我们都尽力打造成大部分时间看起来趋向混乱却有着恰当结构的工作场所：这真的改变了人们关于工作的态度。"